—Geographic
—Information
—Systems
and
Remote
Sensing

Geographic Information Systems
and
Remote Sensing

MORRIS JUPPENLATZ
PhD, SPDip, AADip
Executive Director, Australian Land
Administration Services Pty Ltd

XIAOFENG TIAN
BEng, MEng, GDipIT, DipP
Design Engineer, Optus Vision

McGRAW-HILL BOOK COMPANY Sydney
New York   San Francisco   Auckland   Bogotá
Caracas   Lisbon   London   Madrid   Mexico City
Milan   New Delhi   San Juan   Singapore
Toronto   Kuala Lumpur

National Library of Australia Cataloguing-in-Publication data:
Juppenlatz, Morris.

Geographic information systems and remote sensing: guidelines for use by planners and decision makers.

Includes index.
ISBN 0 07 470327 7

1. Geographic information systems. 2. Regional planning—Remote sensing. 3. City planning—Remote sensing. I Tian, Xiaofeng. II Title.

711.0285

Published in Australia by
McGraw-Hill Book Company Australia Pty Limited
4 Barcoo Street, Roseville NSW 2069, Australia

Publisher: John Rowe
Production Editors: Karin Riederer and Valerie Marlborough
Designer: George Sirett, Asymmetric Typography P/L
Illustrator: Diane Booth

Typeset in 10 pt Caslon
in Australia by Midland Typesetters, Victoria
Printed in Australia by McPherson's Printing Group

Contents

List of figures vii
List of tables viii
About the authors ix
Preface x
Acknowledgments xi

Chapter 1: Concepts and principles

1.1 Information technology 1
   1.1.1 Introduction 1
   1.1.2 The development of the informa-
        tion system 2
1.2 Geographic information 3
   1.2.1 What is geographic information? .. 3
   1.2.2 Gathering geographic
        information 5
1.3 Geographic information systems 7
   1.3.1 Driving force behind a GIS 7
   1.3.2 What is a geographic information
        system? 9
1.4 Remote sensing and geographic infor-
    mation systems 12
   1.4.1 Remote sensing 12
   1.4.2 Integration of geographic infor-
        mation systems and remote
        sensing 16
   1.4.3 Merit and utility of GIS/RS for
        planning and decision making 17

Chapter 2: Components of a GIS

2.1 The data 18
   2.1.1 Database modeling 18
   2.1.2 GIS data types 20
2.2 The software 22
   2.2.1 GIS data collection 23
   2.2.2 Database manipulation 26
   2.2.3 Spatial analysis and interpreting
        capabilities of a GIS 30
   2.2.4 Data presentation (digital
        mapping) 33
   2.2.5 Development tools 35
2.3 Hardware architecture of a GIS 35
   2.3.1 Computer 35
   2.3.2 Input device: digitizer/scanner 38
   2.3.3 The printer/plotter 38
   2.3.4 Storage 39
   2.3.5 Computer configuration and
        network 40
2.4 GIS standards 41
   2.4.1 Data exchange standards 42
   2.4.2 Software standards 43
   2.4.3 Hardware standards 43
   2.4.4 General standardization issues in
        networking 45
2.5 Human resources 45
   2.5.1 Operational staff 46
   2.5.2 Technical professional staff 46
   2.5.3 Management personnel 46
2.6 A simple guide to the evaluation of
    GIS software 47

Chapter 3: Applications of a GIS in planning and decision making

3.1 Principle of using a GIS for planning and decision making 49
3.2 GIS applications in planning and decision making 52
   3.2.1 Local government 52
   3.2.2 Natural resource management 54
   3.2.3 Environmental planning 54
   3.2.4 Emergency planning and management 56
   3.2.5 Mineral resource management 57
   3.2.6 Transport 58
   3.2.7 Public utilities 59
   3.2.8 Socioeconomic development 60
   3.2.9 Private application: marketing research 61

Chapter 4: The approach to acquire a GIS

4.1 Introduction 63
   4.1.1 Approaches for GIS implementation 63
   4.1.2 GIS implementation 64
4.2 Methodology 65
   4.2.1 Theory and tools 65
   4.2.2 GIS technology acquisition management framework 67
4.3 Prefeasibility investigation 68
   4.3.1 Reviewing a range of the organization's objectives 68
   4.3.2 Evaluating the geographic information needs and processing in the business environment 69
   4.3.3 Formulating a long-term GIS project plan and alternative strategies 69
   4.3.4 Selecting preferred GIS strategy   70
4.4 Feasibility study 70
   4.4.1 Project preparation 70
   4.4.2 User requirements analysis 72
   4.4.3 Training issues 78
   4.4.4 Cost–benefit analysis 81
   4.4.5 Revising implementation plan 82
   4.4.6 Preparing a formal proposal 82
4.5 System selection 83
   4.5.1 Revised URAL converted into system requirement 83
   4.5.2 Preselection of suppliers 84
   4.5.3 Preparation of the Request for Tender 84
   4.5.4 Issuing the Request for Tender .. 85
   4.5.5 Tender evaluation 86
   4.5.6 Resolving outstanding issues 87
   4.5.7 Benchmark testing 87
   4.5.8 Final selection 87
4.6 System development 88
   4.6.1 Physical database design and development of operational procedures 88
   4.6.2 The pilot project 89

Chapter 5: Managing the operation of a GIS

5.1 Data capture 90
   5.1.1 Organization issues 90
   5.1.2 Strategy to build a GIS database 9
   5.1.3 Project preparation 9
   5.1.4 In-house data-capture process 9
   5.1.5 Quality issues 9
5.2 GIS data maintenance 9
   5.2.1 GIS database maintenance 9
   5.2.2 Marketing GIS data 10
5.3 GIS technology maintenance 10

Chapter 6: Case studies

Chapter 7: Selected GIS/RS software

Introduction to case studies 102
Case Study 1: GIS for land resources man-
agement and regional
planning (Semarang,
Indonesia) 103
Case Study 2: Pilot Project for Zhuhai
City Government, Guan-
dong Province (People's
Republic of China) 109
Case Study 3: GIS and a Spatial Informa-
tion System for mapping a
human resources profile of
the country (Australian
Bureau of Statistics,
Australia) 117

7.1 GenaMap, with Genasys II 124
7.2 TNTmips, with Microimages Inc. 124
7.3 MicroStation and MGE, with
Intergraph 126
7.4 SPANS, with TYDAC Technologies
Inc. 127
7.5 ARC/INFO with Environmental
Systems Research Institute 129
7.6 Smallworld GIS, with Smallworld
Systems Ltd 130
7.7 MapInfo, with MapInfo Corp. 131
7.8 Mapper, with Earth Resources
Mapping Pty Ltd 132
7.9 IDRISI, with Clark University 133

References 134

Index 137

Figures

1.1 Three dimensions of geographic
information 3
1.2 GIS definition: integration approach ... 10
1.3 GIS and associated information
systems 12
1.4 Wavelength and spectral bands of
emissions from and surrounding the
earth 12
1.5 Remote-sensing satellites in polar
orbit 13
1.6 SPOT satellite descending tracks 14
1.7 SPOT HRV imaging instrument 14
2.1 Component of a GIS 18
2.2 Layers in a GIS 19
2.3 Relational database 19

2.4 Linkage between spatial data and
attribute 20
2.5 Spatial data types 21
2.6 Conversion between data in raster and
vector 22
2.7 GIS data acquisition 24
2.8 Edge-matching processing 25
2.9 Robinson projection 25
2.10 Mercator projection 25
2.11 Miller cylindrical projection 26
2.12 Example of friendly interface 27
2.13 Classification function used in
thematic mapping 29
2.14 Example of SQL function operating
between different databases 30

2.15 GIS data presentation 35
2.16 Components of hardware 36
2.17 Stand-alone model 40
2.18 Centralized model 41
2.19 Distributed model 42
2.20 Configuration of X-Terminals 42
2.21 Network—token ring 44
2.22 Network—FDDI 45
2.23 Open Systems Interconnection model
      of ISO 45
3.1   A GIS in planning practices 51
3.2   GIS applications in planning and
      decision making 52
4.1   Multistage approach 64

4.2   Project life cycle 66
4.3   IT acquisition framework 67
4.4   GIS implementation framework 68
4.5   Organization structure for a GIS
      project 71
4.6   AISIST training modules structure 80
5.1   Spatial and temporal accuracy 97
6.1   Semarang's Integrated Land
      Resources Database 106
6.2   Data processing schema—1996
      Census 121

Tables

1.1   Survey methods 6
1.2   Thematic mapper on Landsat 4,5
      Wavebands and applications 15
1.3   Characteristics of remote-sensed data .. 15
2.1   Attribute and data type 22
2.2   Variables for spatial features
      symbolization 34
3.1   Geographic information systems in
      local government from selected
      countries 53

5.1   Data sources inventory 93
5.2   Map scale versus accuracy 96
6.1   Land resources and regional informa-
      tion needs of the Regional Develop-
      ment Planning Agency of Semarang
      City 107

About the authors

Morris Juppenlatz, formerly Chairman of the Vakgroep VI, Human Settlement Analysis, and Professor of Urban Surveys at the International Institute of Aerospace Surveys and Earth Sciences, Enschede, Netherlands, has been associated with GIS from its early development.

From his PhD in Urban Planning from the University of Edinburgh, his professional activities over 43 years embrace architecture (the Architectural Association School of Architecture, London, in 1951), town planning, urban and environmental planning, urban mapping using small-format aerial photography, land administration and urban land information systems in fourteen different countries. He was admitted as a Fellow of the Royal Geographic Society, London, in 1960, and as a member of the Commonwealth Human Ecology Council in 1971.

Morris Juppenlatz was Director of the Australian Institute of Spatial Information Sciences and Technology, associated with the Land Information Centre, Bathurst, NSW, from 1991 to 1994. Since then he has been Executive Director of the Australian Land Administration Services Pty Ltd, Bathurst.

From 1977 to 1994 he was involved in the transfer of GIS and urban information technology to India (Indian Institute of Remote Sensing, Debra Dun) and to the People's Republic of China (Wuhan Technical University for Surveying and Mapping), where he assisted in inaugurating the Post-Graduate Educational Centre for Urban/Rural Survey, Planning and Management.

His concern, as expressed in this manuscript, is that this new technology can be the means of bringing so many disparate interdependent factors into a computer-controlled synthesis in geographic form for planners and decision making. He regards the modern-day GIS on PCs as the means, within the information technological discipline, of greatly increasing the quality of decision making for society.

The technology has entered the management, decision-making and planning professions so quickly in recent years, that a comprehensible digest to introduce the way this technology can be used for a wide range of applications is now overdue. The authors have attempted to meet this need.

Xiaofeng Tian is currently a design engineer and automated mapping/facilities management specialist in Optus Vision. He holds Dip in Planning, GDip in Information Technology, BEng and MEng.

He started his career as an assistant coordinator for China – European Union Science and Technology Cooperation. In the past decade he has worked as a professional researcher and an engineer across several major areas in which geographic information systems and remote sensing are widely used. Such areas include regional and urban planning, public utility management, resource management, and mapping and land management.

He is one of the software developers for the cost-effective GIS package CARP (Computer Aided Regional Planning). He is an expert in many major commercial GIS packages, such as ARC/INFO, Intergraph MGE, GenaMap, MapInfo and AutoCAD. This skill and knowledge provide him with fundamental understanding and expertise on the application of GIS and remote sensing technology in planning and decision making.

Preface

When the authors joined the Australian Institute of Spatial Information Sciences and Technology in 1991 they came in contact with many local governments throughout Australia. They realized that the planners and decision makers of that broad spectrum of public service were very ill-prepared to select the hardware and software appropriate to their needs, and also there were too few facilities for in-service training of staff. The governing members of the various councils did not know the capabilities of the new geographic information system technology, nor what was required before they made the decision to install a computerized system. In some cases, millions of dollars were expended in the belief that the hardware would solve all their management and planning problems.

The specialist heads of departments at the United Nations Economic and Social Commission for Asia and the Pacific in Bangkok realized that the same conditions prevailed throughout the whole of Asia and the Pacific, wherein nearly half the world's population resides.

The technology and equipment are readily available at a cost. However, the decisions regarding the selection and arrangement of hardware and software, and the appropriate training needed for the staff (most of which could be in-service training), was by no means easy.

The technology, as it has become available to the public service, is among the most advanced scientific knowledge available today, and much of it is user-friendly. More university courses on information technology are being offered, but it is difficult for the busy planner and decision maker to take time off to be instructed in this new technology.

In 1993 Dr He Changchui, Chief of the Space Technology Program at the United Nations Economic and Social Commission for Asia and the Pacific, Bangkok, took the initiative to commission the writing of a manuscript. He envisaged this work would help the planners and decision makers of Asia and the Pacific to understand what is involved in selecting hardware and software for their specific needs.

MORRIS JUPPENLATZ
XIAOFENG TIAN

Acknowledgments

Compiling a manuscript such as this involves inputs from many practitioners from the industry, the suppliers and the users. The authors wish to express their thanks to their many colleagues who have wittingly and unwittingly made a contribution to their text.

In addition, they would like to give a special word of thanks to the highly skilled and experienced staff at the NSW Land Information Centre (LIC). In particular, they would like to thank the Surveyor–General, Don Grant, whose wide vision of the future benefits to be derived from this technology for society as a whole has created such a centre of excellence.

Producing such a manuscript requires much logistical support, and this was willingly provided by the Manager of the Australian Institute of Spatial Information Sciences and Technology, David Mills, and Edwina Brown and Jenny Smith.

The most important acknowledgment of all is to Dr He Changchui, Space Technology Applications Section, UNESCAP, Bangkok, who initiated the request for the compilation of the manuscript.

CHAPTER 1

Concepts and principles

1.1 Information technology

1.1.1 Introduction

In developed countries the economy has been shifting from being material based to being based on information, especially digitally processed information. The new concept is that of an information superhighway. The rapid development of digital and electronic technologies opens a new potential for sophisticated interactive voice, data and image transmission such as digital recording and transmission of sound and pictures, optical fibers for very fast transmission of information, superfast computers, satellite broadcasting and video transmission.

According to recent estimates, more than half of the jobs are already directly or indirectly related to information services, and the figure is likely to grow in the near future. It is increasingly believed that advanced infrastructures for *information exchange and services* will be as dominant in the last decade of the twentieth century as canal, rail and road transport infrastructures were in the previous century. A new industry, *information technology* (IT), has brought about a post-Industrial Revolution that has fundamentally changed business and personal life on this planet.

Information technology is:

> any form of technology, i.e. any equipment or technique, used by people

> to handle information. Mankind has handled information for thousands of years; early technologies included the abacus and printing. The last three decades have seen an amazingly rapid development of information technology, spearheaded by the computer, and more recently, cheap micro-electronics have permitted the diffusion of this technology into almost all aspects of daily life, and an almost inextricable cross-fertilizing and intermingling of its various branches.
>
> DICTIONARY OF COMPUTING 1990

The term 'information technology' was probably coined in the late 1970s to refer to all types of technology and associated resources that relate to the capture, storage, retrieval, transfer, communication or dissemination of information through the use of electronic media. It encompasses all resources required for the implementation of information technology, namely equipment, software, facilities and human resources.

The history of information technology can be traced back to the invention of the computer. Computers were originally used for data processing. Data processing, in turn, was employed largely to ease the large-scale, backroom administrative operations of companies. The nature of the data-processing effort itself

1

has changed from 'data processing' through 'management information' to the more appropriate 'information processing' (Somogyi & Gallier 1987).

1.1.2 The development of the information system

An information system is a key concept in applying information technology. An *information system* is broadly synonymous with the alternative terms 'database system', 'information-processing system' and 'data-processing system'. Essentially it refers to a *system*, with the combination of digital and analog technology, which is designed to input, store, manage, process and output data as meaningful information. The end product, *information*, will normally be a composite entity, constructed from more than one data item.

Typical information systems

Typical information systems include the following (Shelly 1991):

Operational system

An *operational system* is designed to process data generated by the day-to-day business transactions of a company. Examples of operational systems are accounting, billing, banking, inventory control, airline reservation and order entry, along with many other functions. Operational systems are also called *transaction processing*.

Management information system

Computer processing can be used for more than just day-to-day transaction processing. The computer's capability to perform fast calculations, compare data and produce statistical information can be used to produce meaningful information for management. A *management information system* (MIS) refers to a computer-based system that generates timely and accurate information for the top, middle and lower levels of management.

Decision support system

A *decision support system* (DSS) is a system designed to help someone reach a decision by summarizing or comparing data from either internal or external sources or both of these sources. Internal sources include data from an organization's files, such as sales, manufacturing, services or financial data. Data from external sources could include information on interest rates, population trends and market trends. Decision support systems often include query functions, statistical analysis capabilities, spreadsheets, graphics and mapping functions to help the end user evaluate the decision data. More advanced decision support systems also include capabilities that allow end users to create a model of the factors affecting a decision.

Expert system

An *expert system* combines the knowledge on a given subject of one or more human experts into a computerized system that simulates the human experts' reasoning and decision-making processes. Thus, the computer might also be considered to possess expert knowledge on the subject. Expert systems are made up of the combined subject knowledge of the human experts, called the *knowledge base*, and the *inference rules* that determine how the knowledge is used to reach decisions.

Integrated information system

With today's sophisticated software, it may be difficult to classify a system as belonging exclusively to one of the abovementioned information systems. An *integrated information system* will perform all the functions from operational to management and finally to decision making.

Application of an information system

The application of an information system can be found in most industries in developed countries, for example in national/state/local government, financial institutions, wholesalers/retailers, transport, utilities, manufacture, communication and various service industries.

In the fields of planning and decision making, IT and information systems offer almost unlimited opportunities to improve understanding and practice, to develop new theories, and to increase the effectiveness of urban and regional planning. A series of new tools, such as statistics analysis packages, census data-processing systems, customized management and decision support systems, expert systems and geographic information systems, are being widely used in various levels of the planning and decision-making processes.

1.2 Geographic information

Effective formulation and implementation of any strategies, policies and plans are highly dependent on accurate, comprehensive and timely information. As most issues are interdisciplinary, planners and policy makers require access to sectoral data, including economic, demographic, social, geographic, environmental and natural resources data. The data processing includes the data collection, manipulation, and representation. Geographic data is now identified clearly as that required for geographic information systems, as against data for demographic needs, economic statistics and so on.

Many researchers claim that between 75% and 90% of the information used every day by most organizations is geographically based. For planners and decision makers, geographic information is especially important. The *geographic information system* (GIS) is one of the most powerful tools in planning and decision making today.

1.2.1 What is geographic information?

Data and information are abstracted to represent the real world and the events in it. Data and information can be considered as having three different dimensions: theme, time and location (Salomonsson 1980). When the information or data is referenced to the objects or the events that are geographic features on earth, then the information is *geographic information*.

Three dimensions of information
Geographic data and information have components of spatial data, attribute data and time (Fig. 1.1).

Spatial data
Spatial data is the data pertaining to the locational aspect of geographical features together with their spatial dimensions. They are approximated by point, line and areal extent.

Point
At the global scale a point may be used to define the location of a metropolis.

When dealing with urban planning issues, a point would not be enough to define a city, but it may be suitable to describe the location of a school or hospital within the city.

Line
At the global scale a line of coordinates may be suitable to define the linear extent of a river, road and so on.

When dealing with local issues, a road could be represented as an area object, together with a property boundary at a very large scale. In this case, the utilities network could be represented by lines.

All conceptual line objects can be represented by line, such as segments of an administrative boundary, segments of a legal boundary of the road reserve and contours.

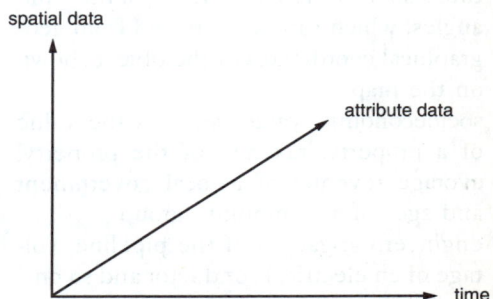

Fig. 1.1 *Three dimensions of geographic information*

Areal extent

At the global scale a closed line of coordinates may be enough to define the area of an island, a country or a desert.

When dealing with regional issues, it may be necessary to define species and subspecies within the forest, sensitive soil and so on, by using a set of closed line coordinates.

Attribute data

Attributes are the description, measurement and classification of the geographic features. Attributes are presumed to be identical to the whole geographic feature. A geographic feature could have many attributes depending on the degree of significance. Generally speaking, attribute data has descriptive, quantitative and qualitative aspects.

Descriptive data does not have a quantitative or qualitative nature. Elements of the data cannot be compared with each other in quality and quantity. Examples of descriptive character are:

- property ID number
- property owner's name
- facility types
- road-building material types
- vegetation types
- zoning area types

Attribute data that can be distinguished by measurement and numbers is *quantitative data*. Examples are:

- dimensional: measurement of spatial features such as length, area, volume and angles, which can be extracted from geographical coordinates of the object shown on the map
- socioeconomic: values such as the value of a property, tax rate of the property, average revenue of a local government and ages of a community group
- engineering: gauges of the pipeline, voltage of an electrical conductor and so on

Attribute data can be *qualitative data*, such as the following:

- water
- air
- soil quality
- steepness

Themes, one aspect of attributes, can be used to identify all spatial features of the world that are related to a particular feature or theme, hence the term 'thematic attributes'. Each of these features, which can be considered as themes in themselves, can be further divided into subthemes. For example, waterbodies may include:

- oceans
- seas
- lagoons
- lakes
- springs
- dams

Time

Geographic information can change over time. You notice this when you cannot find a road on a map drawn a few years ago. Time is another factor that can be included in a geographic information system to cope with the problem of:

- monitoring change in attributes, for example a change in river dimension and location with flooding in a specific time frame, and changes in the use of land from agricultural to industrial over specific time periods
- monitoring change in locations of objects, such as navigating each individual real-time movement of a fleet within the operation area, and tracking the living pattern of a certain kind of wildlife.

Typical geographic information

Generally, geographic information can be classified as regional or thematic (topical). *Regional geographic information* includes more than one kind of geographic information in a place or region, while *thematic geographic information* may be the one earth feature or human activity as it occurs throughout wide areas. Thematic geographic information may include physical and human geographic information.

Physical geographic information

Physical geographic information concerns the locations of such earth features as land, water and climate, and their relation to one another and to human activities. Physical features include:

- landforms
- climatic patterns
- vegetation classification
- animal distribution
- soil types

Human geographic information

Human geographic information concerns patterns of human activity and their relationships with the environment, such as:

- *political:* the political boundaries, pattern of voting and political instability
- *cultural:* beliefs, customs and other cultural traits
- *population:* birth/death rates, population movements, family size and other statistical data
- *social:* employment, professions, welfare, social bias and criminal pattern
- *economic:* growth rate, taxation, income, trade, industrial structure and production
- *historical:* the places and patterns of human activity that change over time
- *constructed:*
  - buildings, and other kinds of constructed structures, such as dams and parks
  - roads, railways, waterways, and pavements for transport
  - utilities and facilities: electricity, gas, water, sewerage, stormwater, communication, and so on

1.2.2 Gathering geographic information

There are well-established theories and methods of geographic information collection, management, analysis and visualization following thousands of years of development. These are known as *survey and mapping/cartography.*

The source of geographic information

Originally, all geographic information data came from *survey*, and society has accumulated geographic information through hundreds of years of survey practice. In this text the concept of survey is not limited to field survey or socio-economic survey—it is the general definition of the measurement of the spatial data, attribute and time of geographic features. The most common methods of survey involve fixing one of the components, controlling another and measuring the third. There are several survey methods: field, aerial, census, statistics and tracking (see Table 1.1).

Field survey

Field/hydrographic/mining survey is the surveying of the land, underwater areas and underground features of the earth. There are several kinds of survey services such as the following:

- Geodectic survey is the most accurate or precise form of control survey whose main purpose is to provide the master framework within which other types of survey of lesser accuracy can be located.
- Topographical survey is concerned with the measurement of natural and artificial features on the earth's surface so that a map of these features may be produced. Frequently, this type of survey is now carried out by means of aerial photography, and the term 'topographical survey' tends to be used to refer to second and lower orders of survey control.
- Cadastral survey is concerned with defining, demarcating, measuring and recording the exact boundaries of properties.
- Engineering survey is a survey made specifically to supply particular information for engineering purposes.

Aerial survey/Remote sensing

Aerial surveys utilize photographs taken from an aircraft. Measurements are obtained from the photographs or from three-dimensional projections of stereo pairs of photographs. The discipline that deals with these measurements

Table 1.1 *Survey methods*

Kind of survey	Fixed	Controlled	Measured
Field survey	Time	Attribute	Spatial
Aerial survey/RS	Time	Spatial	Attribute
Census	Time	Spatial	Attribute
Statistics	Spatial	Time	Attribute
Tracking	Attribute	Time	Spatial

is called *photogrammetry*. It is the art, science and technology of obtaining reliable information about physical objects and the land environment through the processes of recording, measuring and interpreting photographic images, and analyzing patterns of recorded radiant electromagnetic energy and other phenomena.

Remote Sensing (RS) is the measurement of visible light and electromagnetic radiation reflected or emitted from the surface of the earth and its surrounding atmosphere. These measurements are usually taken from a platform on a satellite or aircraft well above the surface of the earth. The remote-sensed images are transformed into usable information through image analysis procedures. (See Fig. 1.4, p. 12.)

Census

Census and social survey is a survey usually conducted by a national government organization to gather information about the society that it governs. It may not be limited to a population census, which determines the size of a population, but may cover such information as the age, sex, race, employment and income of people. Other censuses examine such subjects as housing, agriculture and industry.

Statistics

Statistics is a set of methods used to collect and analyze data. It includes the collection and study of data at different time intervals and at a fixed location, to provide information for yearbooks, weather station reports, and so on.

Tracking

Tracking/monitoring is carried out to obtain the information on changes that occur at a location over several periods of time. Examples are monitoring the change of the ecosystem, and a real-time monitoring of a moving object such as a vehicle.

Mapping: The manipulation of geographic information

Both words and numbers have been used to describe geographic information since language was invented. But alphanumeric descriptions of spatial data were abstract and inadequate, and were not actually describing the physical representation of the geographic feature itself.

The most common medium for storing and displaying such coordinate-based information has traditionally been the *analog map document*: 'a map is worth thousands of words'. A map is a subjective record of a portion of the earth's surface. Information portrayed on a map has been selected, simplified, classified and then symbolized. The information is not a complete record of what exists in an area, but consists of selected features portrayed in such a way that the user of the map gains an impression of the mapped area.

A map is produced through mapping technology, which is the analysis and graphic representation of the geographic information. Geographic information has been stored on base maps of paper, vellum and other media for at least 2000 years. A map contains:

- a common spatial reference to define the location of identities
- symbols of point, line and area, and the combination of these to characterize *geographic identities* at reduced scale
- colours, pattern and other properties of symbols to customize the basic symbols

Spatial data elements are recorded on maps as points, lines and areas within the basis of a common, three-dimensional standard coordinate system (such as latitude, longitude and elevation with respect to sea level). The major functions of maps are normally the identification of spatial data elements, the determination of their locations, the measurement of their spatial attributes, and the subsequent storage and portrayal of these data elements on maps. Retrieval and analysis of this map data normally involves visual inspection of the map document coupled with intuitive analysis, which is occasionally aided by simple measurement tools.

The production of the first accurate base maps was significant in mapping. It can be dated back to the mid-eighteenth century. *Thematic mapping* (Parent & Church 1987) is defined as a representation of attributes to a given geographic area. Such maps portraying magnetic variation with isoclines, wind direction (by means of arrows) and other tactical information began to appear after the mid-eighteenth century. The idea of recording different layers of data on a series of similar base maps was born in the United States during the American revolutionary war. The technique of *hinged overlays* was invented by French cartographer Louis Alexander Berthier. The refinement of lithographic techniques enabled the thematic cartographer to become more precise in data representation.

By 1835 cartographic techniques, social science theory and environmental awareness had advanced to the point where the combination of these three factors could support more comprehensive thematic mapping projects in Great Britain (Parent & Church 1987). With the Industrial Revolution came the rapid expansion of the manufacturing industries, and this drew people into the already crowded urban settlements. It was not until the latter part of the reign of Queen Victoria that the need for an extensive urban infrastructure was recognized. The atlas to accompany the second Report of the Irish Railway Commissioners, published in 1838, consists of a series of maps that depict population, traffic flows, geology and topography. For each sheet, the base map was uniform in regard to scale and county boundaries. By manually overlaying the different attributes at a given spatial location, the commissioners could make their recommendations as to where the best transportation routes could be sited.

Thematic mapping continued to improve with the London publication in 1848 of the first truly uniform world atlas in respect to scale, symbolization and cohesiveness. Such conventions as isobars, graduated circles and volume lines were used extensively.

The cartographic techniques continued for 100 years. It was not until the development of electrical computers, about 1952, that the ENIAC (Electronic Numerical Integrator and Calculator) computer began processing the 1950 Census data. The process of electronic calculation opened new possibilities of research based on the massive manipulation of huge data files. Intricate models and alternatives could be generated to simulate future events.

1.3 Geographic information systems

1.3.1 Driving force behind a GIS

It is not difficult to retrieve small amounts of data manually from a single map, but the retrieval of larger numbers of map elements from one or more maps is a very slow process.

With the introduction of the 1947 *Town and Country Planning Act* in Great Britain, a new cadre of town planners were trained to undertake land-use planning. This change in emphasis of the planning technique required the town planner to undertake the statutory planning directly related to the reality of the land, that is, land use related to the cadastral plan, the actual owner of the land, the occupier of the land and uses of the individual parcels of land.

It is a very skilled task to obtain an accurate measurement, such as distance and area, when quantitative analysis is required. Maps have limited capabilities in spatial analysis, such as overlay, surface analysis and network analysis.

Maps show little information beyond location, coordinates and themes.

The high cost and slow operation associated with the manual compilation, drafting, printing, distribution and storage of paper or cloth maps make the *manual mapping system* inefficient, and also limits the provision of up-to-date information.

On the one hand, maps are made to meet general needs. However, it is very difficult to produce a map to meet the individual needs of eveyone. On the other hand, information on user-tailored maps has little use to any other users. Much geographic information accumulated in the past is now shown to be redundant and segregated.

Maps are a static representation of a moment in the physical world within a short time. But many applications require effective temporal-spatial modeling tools to monitor the movement and the change of geographic features. Traditional maps are limited to the recording of static geographic information and lack the representation of dynamic functions. Areas that are large relative to map scale will need to be represented by several connecting maps. Care always has to be taken to ensure that the information at the edge of adjoining maps actually matches.

Information technology enhances the collection, management, analysis and representation of geographic information to attain an information system, or a *database management system* (*DBMS*). The DBMS is a kind of software system designed for management purposes, and it provides facilities to organize a database and allow shared access and control of the database. It also maintains the reliability, security and integrity of the data (Oxborrow 1986). Even a simple wordprocessor and spreadsheet program can be used to store, manage, manipulate and analyze attribute data of geographic information. Special packages, such as statistical analysis packages, are used for comprehensive attribute data analysis.

Digital mapping technology automates the traditional map-making process. The improvement in one or two aspects of geographic information processing can solve only some of the problems. The concept of a *geographic information system* that would integrate all requirements of the database management, spatial analysis and digital mapping requirements has emerged in the past decades.

The history of the GIS has its beginning in the 1960s. In 1962 Roger Tomlinson of the Canada Land Inventory developed the Canadian Geographic Information System (CGIS) (Peuquet 1977), the first system to be called a 'geographic information system'. It was designed for more than just one specific application. Its major application was to store digitized map data and land-based attributes in an easily accessible format for all of Canada. A polygon-based system, it had significant shortcomings in interactive capabilities to the point that real-time graphic editing was severely limited. The CGIS had storage and retrieval capabilities, plus it could reclassify attributes, change scales, merge and create new polygons, and create lists and reports. The Canadian Geographic Information System was implemented in 1964 (Deuker 1979), one year after the first conference on Urban Planning Information Systems and Programs. This conference led to the establishment of the Urban and Regional Information Systems Association.

The New York Landuse and Natural Resources Information System was implemented in 1967, and the Minnesota Land Management Information System was implemented in 1969. In these early years the costs and technical difficulties of implementing full-scale geographic information systems were such that only large users of geographic information, such as federal and state agencies, could afford their development.

In addition to the beginning of commercial GIS development, the 1970s also saw significant developments in *remote sensing* and *image processing systems*. Developments in remote sensing technology and application during the 1970s spurred practical and theoretical work in the areas of geometrical corrections and registration.

There has been a rapid increase in the number of GIS packages that have become available, as a result of both advances in computer technology and increases in the availability of spatially referenced data in digital form. A United States Department of the Interior, Fish and Wildlife Service Report, published in 1977, compared the selected operational capabilities of fifty-four different geographic information systems in the United States. Geographic Information Systems were proven to assist in the management, analysis and presentation of information in a suitable graphic form for planning and decision making.

This new digital mapping process was first proven to be of special importance in environmental and natural resource information management, especially in developing countries. World Bank and other United Nations (UN) agencies, such as United Nations Environmental Program (UNEP) and United Nations Economic and Social Commission for Asia and Pacific (UNESCAP), are the leading international development agencies in the introduction and use of GIS for the promotion of more effective approaches to natural resource development planning and management.

Geographic information systems and remote sensing technology have become the core management system of some sections of the United Nations Development Program (UNDP) and other international fund agencies. These agencies fund projects in regional economic planning and development, environmental protection and natural resource management. Through various international and bilateral funding, the application of parts of GIS technology can now be found in some developing countries throughout Asia and the Pacific for planning and management purposes.

To date, there are a variety of commercial geographic information systems available in the world that are continually expanding their capabilities (see Chapter 7). There are also many attempts that have failed and noncommercial geographic information systems that existed in the past but have not progressed. A GIS industry with billions of dollars turnover annually, which consists of software, hardware, data and information consultancy, training, education and so on, has now been established.

1.3.2 What is a geographic information system?

It is not easy to give a geographical information system an unarguable definition because it associates with so many disciplines, such as information technology, system engineering, survey and photogrammetry, cartography, engineering, socioeconomy and geography. The following is a list of approaches to defining the GIS (AISIST 1993c).

Systems approach

A GIS has four functional components (Marble & Amundson 1988):

1. A data input subsystem collects and/or processes spatial data derived from sources, such as existing maps, remote-sensed data and direct digital input.
2. A data storage and retrieval subsystem organizes spatial data in a topologically structured form, which permits it to be quickly retrieved on the basis of either spatial or nonspatial queries for subsequent manipulation, analysis or display. This subsystem must also permit rapid and accurate updates to be made to the spatial database.
3. A data manipulation and analysis subsystem performs a number of tasks, such as changing the form of the data through user-defined aggregation rules, or producing estimates of parameters for transfer to external analytical type models.
4. A data-reporting subsystem is capable of displaying all or selected portions of the spatial database in terms of standard reports or in a variety of cartographic formats.

Transformational approach

A GIS transforms data into information by integrating different *data sets*, applying focused analysis and providing output, all in a manner designed to support decision making.

Integration approach

A GIS integrates data sets that were previously difficult to merge or overlay, allowing for the establishment of a more complete picture of the problem at hand. This approach sets a GIS as an integrating tool more than an analytical tool. Refer to Figure 1.2.

Modeling approach

A GIS integrates spatial and nonspatial data in two, three or four (for example, time series) dimensions. This allows for complex modeling of data to solve problems. This approach sees a GIS primarily as an analytical tool.

Spatial boundaries

Attributes

Socioeconomic
- census district
- administrative units

- population
- income

Facilities
- telephone
- water
- electricity

- size
- material
- last inspection date

Land administration
- land parcels
- easements

- owner
- value
- use

Natural resources
- vegetation
- water
- agriculture

- vegetation type
- land cover

Topography
- contours
- rivers
- roads

- heights
- place names

Terrain
- 3-D model

- height
- slope
- aspect

Common spatial referencing framework

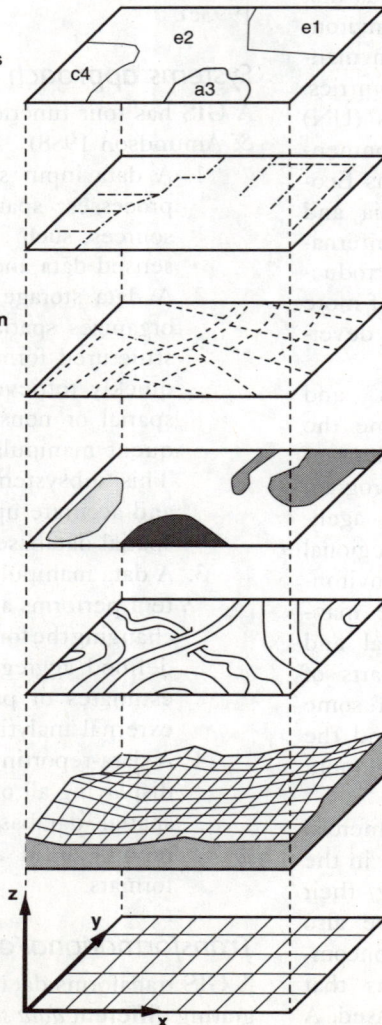

Fig. 1.2 *GIS definition: integration approach*   AISIST 1993b

The real-world definition of a GIS

The definitions of a GIS given above and in the various texts are very broad. In practical application, a GIS covers a broad spectrum of functionality, ranging from a simple computer mapping system, which does little more than overlay various maps, to a full analytical system able to solve complex spatial problems in three or four dimensions.

To summarize all the definitions mentioned above:

- A GIS as an *information technology*, in the first instance, is a system of ordering, managing and accessing large quantities of information. This task corresponds, in part, to that of nonspatial information systems. The difference, however, lies first and foremost in the additional opportunity of integrating various data sources, as all information has a component of geographical or spatial position, which is the common element that allows the integration of data from different sources.

- A GIS not only is an information system for data management but also is a system that allows analyses, which specifically relate to the spatial or land component of the data, to be carried out.

- A GIS is capable of providing a solid cartographic presentation. Maps need no longer be drawn manually, but can now be produced from a digital mapping system contained in the GIS.

- A GIS is not designed and installed solely for the purpose of being a GIS, but for its *applications* to several needs of planning, management decision making and operational needs of government and the private sector. A GIS should be incorporated strongly with applications. Commercially available GIS software usually requires further system development to accommodate specific local needs.

- A GIS is *not* just an information technology. A GIS incorporates technology, people, the data, the organization and skilled personnel to make it an effective aid to management and decision making.

The technology itself will not prove effective if data needs, human resource management and organization change are not addressed. Failure in GIS implementations is often related to a lack of understanding of the broader issues.

In some cases users substitute another word or expression for 'geographic' to give a more precise meaning to the purpose of the information system, such as the urban information system or the natural resource information system (Carter 1988). On the other hand, there are other types of geographic information that are so distinct from the generic GIS that the personnel working in these systems tend to operate in their own communites of interest. The automated mapping/ facilities management (AM/FM) community is primarily concerned with utilities and the built environment. This community now has its own organization and literature. Another community that might be thought of as standing apart from the generic GIS is that community focused on the land parcel or, as it is more commonly referred to in the international sphere, the *cadastre*. Many persons now refer to such a system as a land information system (LIS), of which the Digital Cadastral Data Base provides the exact location on the ground for identifying specific land/plot information. For some professionals the scale of investigation and the specific focus of the LIS community are distinct enough to stand apart from the more generic GIS (Fig. 1.3).

There are also many systems that are functionally geographic information systems but that are not thought of as geographic information systems. Among these systems are weather forecasting systems, marine exploration and navigation systems, petroleum exploration systems, and transportation routing and modeling systems. There is no universal geographic information system, nor should we ever

| spatial information system | |
| land information system | |

| desktop mapping/ analysis system | |
| automatic mapping facility management system | |

| urban information system | |
| natural resources management system | |

| environment planning system | |
| emergency management system | |

| spatial decision support system | |
| mineral exploration information system | |

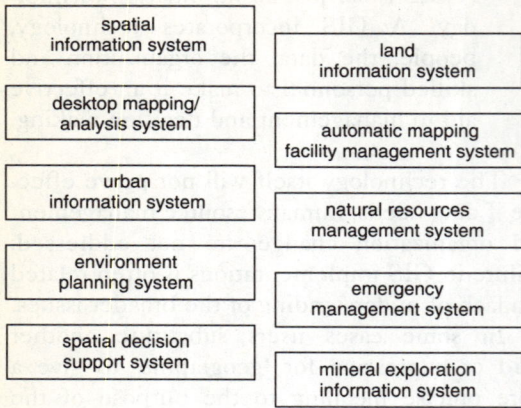

Fig. 1.3 *GIS and associated information systems*
AISIST 1993c

expect to see one, for there is little communality between the variety of tasks that such systems are designed to address. However, these systems have at least one thing in common—each deals with data and information that is georeferenced. Geographic information systems available are usually organized around functional tasks.

1.4 Remote sensing and geographic information systems

1.4.1 Remote sensing

Remote sensing concepts

Remote sensing is a technology that collects data relating to the earth's surface without contacting with it, through a sensor mounted in a satellite or high-flying aircraft.

The earth's surface and atmosphere emit individual characteristic signatures within the visible light and electromagnetic radiation spectrum. The spectrum is divided into spectral bands ranging from short gamma rays to long radio waves (Fig. 1.4).

Visible and infrared light, along with electromagnetic radiation from any object, can be absorbed, transmitted, reflected or scattered

once it contacts with other objects. The success of remote-sensing operations depends upon the technology and knowledge of interpretation of the effect of this variation to the radiation. In order to record this wide range of wavelengths, three main types of sensors are commonly utilized.

Photographic systems

Photographic systems, utilizing sophisticated camera equipment and film, are one of the most widely used and commonplace remote-sensing systems. Within this type of system, the film is used to record the visible and/or near-infrared (NIR) radiation detected by the film in the camera at the time of exposure. True color films are commonly utilized to record the visible end of the spectrum, while monochrome, infrared (IR) and false color infrared film are typically

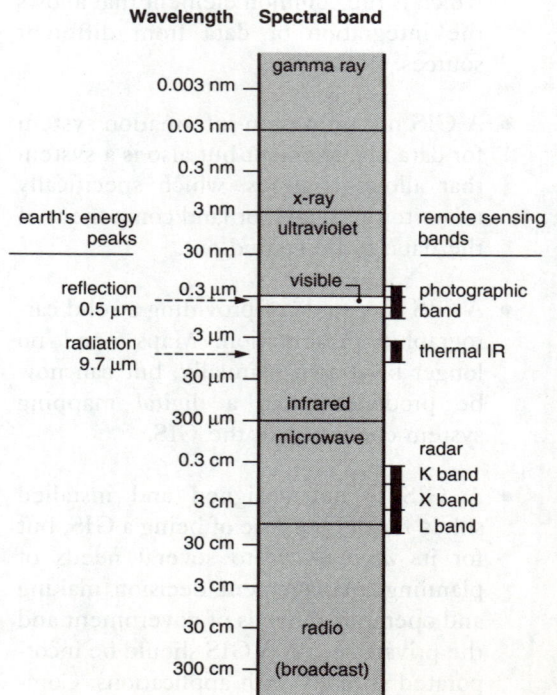

Wavelength	Spectral band	
0.003 nm	gamma ray	
0.03 nm		
0.3 nm		
3 nm	x-ray	remote sensing bands
30 nm	ultraviolet	
reflection 0.5 μm — — 0.3 μm — —	visible	photographic band
radiation 9.7 μm — — 3 μm — —		thermal IR
30 μm	infrared	
300 μm		
0.3 cm	microwave	radar
3 cm		K band
30 cm		X band
3 cm		L band
30 cm		
30 cm	radio	
300 cm	(broadcast)	

earth's energy peaks

Fig. 1.4 *Wavelength and spectral bands of emissions from and surrounding the earth*   RAJAN 1991, P. 3

used for the near-infrared part of the spectrum, especially to record water bodies.

Linescan systems

Linescan systems involve the use of an oscillating mirror placed in parallel to the direction of the satellite or airplane motion. Radiation, perpendicular to the direction of the flight, is reflected by the mirror and received by an on-board detector. Scan lines are then recorded in sequence across the area of interest.

The most common application of such systems is within the visible and near-infrared portion of the spectrum, although the systems can be used to record all wavebands, including microwave, if equipped with an appropriate detector.

Radar systems

Radar systems normally emit their own radiation and detect the radiation back from the earth's surface. The radar receiver can collect and record the active energy or pulse return to the sensor.

Platforms

The platforms on which sensors and receivers are mounted are separated into low-altitude aircraft and high-altitude satellites and spaceships.

Satellites are the common carriers of such systems. Since 1972 many such satellites have been launched into polar orbit, which is about 900 kilometres (560 miles) from the earth, by the United States, France, former USSR, Japan, China, India and so on. It takes around 100 minutes for each satellite to complete each orbit, and around 16 days to return to its starting point and repeat the cycle (Fig. 1.5).

Remotely sensed data

The Landsat Series (US), and SPOT series (France) satellites have provided users around the world with versatile, high-resolution, multispectral data, together with several satellites from China, India, Japan and so on. Landsat 4 and 5 use thematic mapper (TM) to register images in seven spectral ranges, with 185 kilometres (115 miles) width of image across the earth's surface. There are 233 fixed orbital paths spaced equally around the earth, and each path is covered once in 16 days. (Rajan 1991). See Table 1.2 and Figure 1.6.

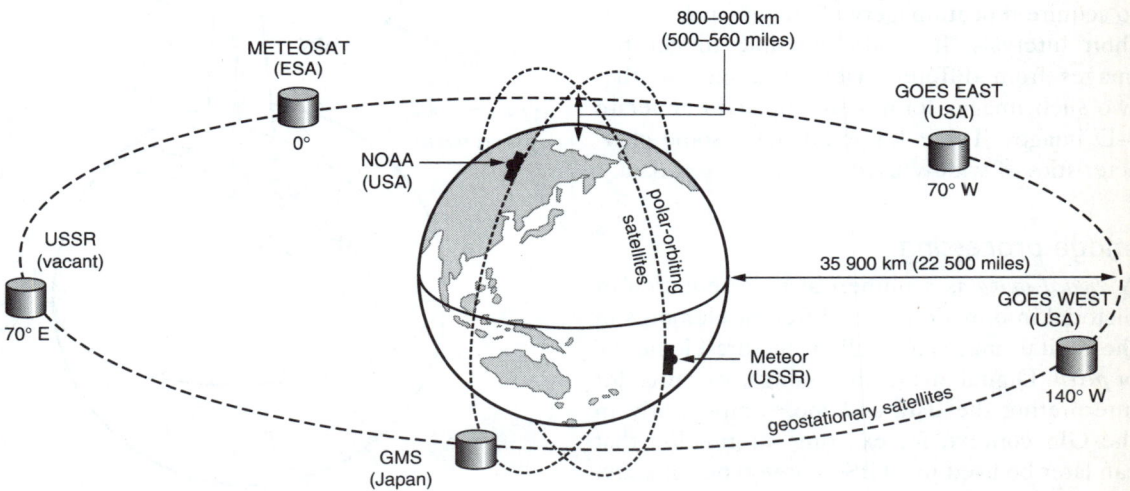

Fig. 1.5 *Remote-sensing satellites in polar orbit, compared with communication and some weather satellites in geostationary orbit, which is about 35 900 km (22 500 miles) away from the earth's equator* RAJAN 1991, P. 7

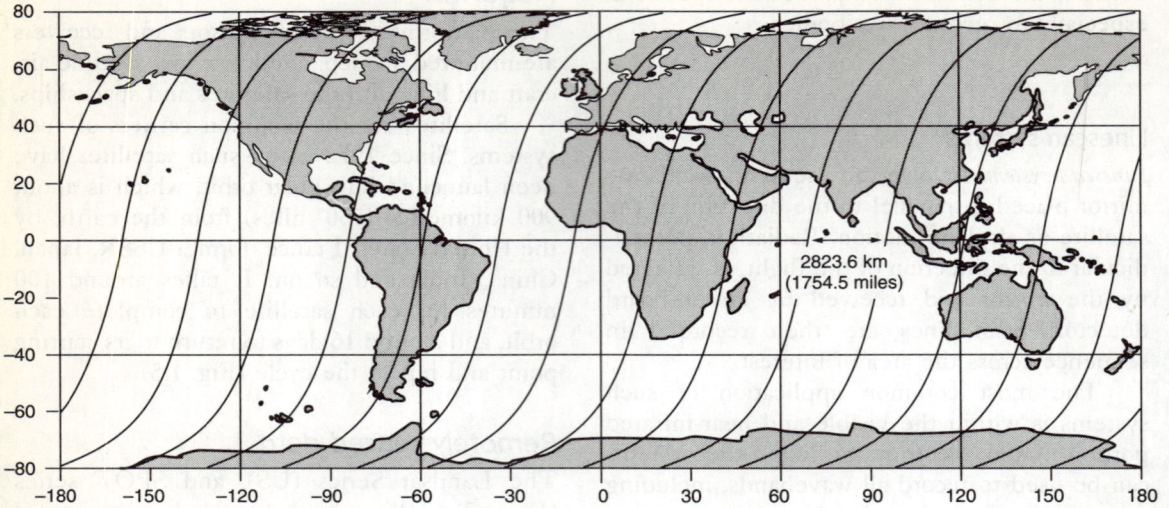

Fig. 1.6 *SPOT satellite descending tracks on a given day* CURRAN 1985, FIG. 5.23, P. 157

SPOT has two identical high-resolution visible (HRV) imaging instruments, each of which acquires images of a strip 60 kilometres (37 miles) wide. The combined field of view is 117 kilometres (73 miles) with 3 kilometres (1.8 miles) overlap (Fig 1.7). It acquires image data in two bands in the visible part (green and red) and one in the near-infrared. The oblique viewing capability of SPOT makes it possible to acquire repeat imagery of any given region at short intervals. It is also possible to get the images from different viewing angles so that two such images form a stereopair to generate 3-D images. Table 1.3 summarizes some characteristics of widely used remote-sensed data.

Image processing

A *digital image* is a numerical representation of pictorial information. The different elements in the digital images are called 'picture elements' or *pixels*. Digital image processing is essential for interpreting the data and more importantly in the GIS context, for exacting information that can later be used in a GIS. Three types of activities are involved:

- *Preprocessing of the image* is carried out when the raw data has been obtained.

Primarily, the data should be calibrated, typically through radiometric calibration. Secondly, a geometric correction should be applied to the image.

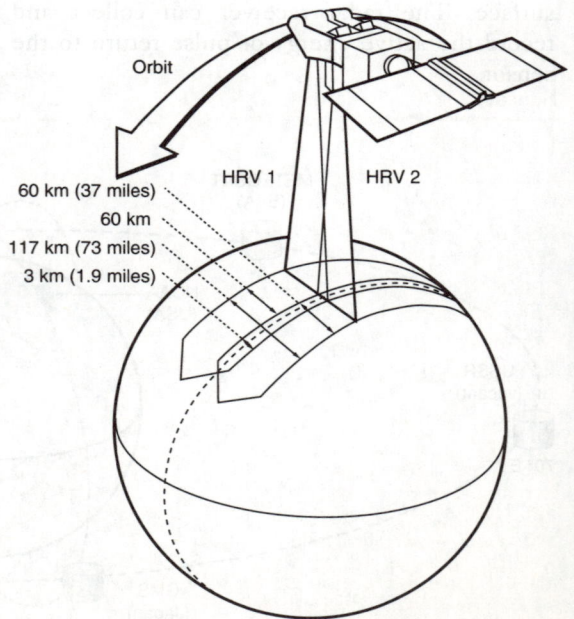

Fig. 1.7 *SPOT HRV imaging instrument* RAJAN 1991, P. 27

Table 1.2 *Thematic mapper on Landsat 4, 5: Wavebands and applications*

Band	Spectral range (um)	Features and applications
Blue	0.45–0.52	Good water penetration, strong chlorophyll absorption. Mapping of coastal water areas. Differentiation between soil and vegetation. Differentiation between coniferous and deciduous vegetation.
Green	0.52–0.60	Matches green reflectance peak of healthy vegetation, sensing the health of vegetation.
Red	0.63–0.69	Chlorophyll absorption band, very strong vegetation absorption. Differentiation between plant species thanks to the chlorophyll-absorption assessment.
Near-IR	0.76–0.90	Complete absorption by water. High land/water contrasts, very strong vegetation reflectance. Survey water body delineation.
Near-middle IR	1.55–1.75	Very moisture sensitive. Differentiation between clouds and snow cover. Measurement of vegetation moisture and soil moisture. Reflectance of most rock surfaces.
Thermal IR	10.4–12.5	Thermal imaging and mapping. Information on plant heat stress. Thermal data on geological information.
Middle IR	2.08–2.35	Good geological discrimination. Hydrothermal mapping. Rock-type discriminations (mineral and petroleum).

- *Image enhancement* involves the improvement of the visual capabilities of the image through a number of methods.

- *Image classification* involves the categorization of pixels into like groups to produce a thematic map.

Table 1.3 *Characteristics of remote-sensed data*

Satellite/ Sensors	Country	Resolution (m)	Swath width (km)	Return pass (days)	Electromagnetic ranges
Landsat/MSS	USA	80	180	16	Green, Red, NIR
Landsat/TM	USA	30	180	16	Blue, Green, Red, NIR, NMIR, Thermal
SPOT/ Panchromatic	France	10	60	26	Green, Red, Panchromatic
SPOT Multispectral	France	20	60	26	Green, Red, NIR
NOAA/AVHRR	USA	1100	>2500	Daily	Red, NIR, Thermal
MOS/MESSR	Japan	50	100	17	Green, Red, NIR
MOS/VTIR	Japan	900–2700	1500	17	Panchromatic, Thermal
MOS/MSR	Japan	NA	317	17	Microwave
IRS	India	36.5	148	22	Blue, Green, Red, NIR

MSS: Multispectral scanner
TM: Thematic mapper
AVHRR: Advanced very high resolution radiometer
MESSR: Multispectral electronic self-scanning radiometers
VTIR: Visible and thermal infrared radiometer
MSR: Microwave scanning radiometer; NIR: Near-infrared; NMIR: Near-middle infrared
SPOT: Système Probatoire d' Observation de la Terre (France)
NOAA: National Oceanic and Atmospheric Administration (USA)
MOS: Marine Observation Satellite (Japan)
IRS: Indian Remote Sensing (India)

1.4.2 Integration of geographic information systems and remote sensing

Geographic information systems and remote sensing have been developed from different scientific disciplines and application areas. There has been a long-term interest in merging these two technologies. The differences between the image processing of remote-sensed data and a geographic information system blur when remote-sensed and other geographically registered data are merged. GIS and remote-sensing technology are becoming an integrated technology that is being widely used in various applications. Several of the GIS suppliers, such as the Environmental Systems Research Institute (ESRI) and Intergraph, now include image-processing functionalities for processing remote-sensing data from Landsat and SPOT, by which the raster images can be integrated with the vector details of the same area of land.

The similarity between RS image processing and GIS techniques lies in the fact that both have to deal with spatial data, and both can be mapped digitally.

In one sense, both remote sensing and GIS are complementary, as they are simply variants of *digital spatial data*. From the GIS point of view, remote sensing is one of the cost-effective and important data acquisition technologies. The integration of remote-sensing information into a raster-based GIS can occur easily, because the base data structures are similar. Remote-sensed imagery can be efficiently stored in raster structures within a GIS without compromising spatial resolution or accuracy.

The spatial display techniques for geographic and remote-sensing information has greatly advanced within the past two decades. Technological advances in image processing and visualization techniques have developed display and interpretation mechanisms for the analysis of all sources of geographic information. Once the remote-sensed image is compatible with the GIS, a variety of analytical functions can be applied to the remote-sensed data.

It should be noted that the largest effective scale for detailed analysis of SPOT or Landsat images is 1:15 000. For detailed planning analysis, it is appropriate to use Small Format Aerial Photography (SFAP) over the built-up areas that are under study.

Thus, remote sensing and geographic information systems are becoming inextricably linked in many application areas. *Natural resource* managers are increasing their use of remote-sensed imagery in combination with vectorial GIS data, such as cadastral boundaries, elevation models, and infrastructure to improve mapping of resources and modeling of economic, social and environmental processes. Urban and regional planners are also using processed remote-sensed data, sometimes in three dimensions, as background to display cadastral boundaries and infrastructure data, and improve planning and monitoring of existing plans. Even in the area of the provision of utilities, demand is being assessed and new services planned, by using remote-sensed imagery and GIS data together.

Spatial software models, in addition to mainstream suppliers, are now becoming available for assessing and analyzing remote sensing data in conjunction with other geographical data sources. These packages are becoming more user friendly, and speed up the process of integrating the data from both sources. Some of the major GIS vendors have developed modules for image processing that could be run as an integrated system. The successful integration of GIS and remote-sensing systems require not only a combination of the standard functionality of both a GIS and remote sensing, but also the active combination of attributes across all data sets, regardless of their origin, and the subsequent analysis of the data within a multidimensional environment.

The skills involved in such integration are beyond the normal skills required for digital capture and GIS manipulation. Specialist training will be required for the staff to perform the processing, enhancement and image classification.

1.4.3 Merit and utility of GIS/RS for planning and decision making

The inventory, analysis, mapping and modeling capabilities of a GIS have wide applications in a range of planning practices, ranging from data retrieval and site selection to project monitoring, development control and programming.

> A GIS is most useful at the analysis, forecasting and plan formulation stages. Through the use of map overlays, problem areas, such as urban redevelopment districts and environmental sensitive areas, can be identified. At the plan formulation stage, land suitability maps can be generated using a GIS for the preparation of master plans and land-use plans. It can also be used to consider environmental aspects of zoning, and to determine areas which are unsuitable for residential development, owing to the presence of polluting industries, hazardous environments or traffic noise. With the use of a GIS, the spatial impacts of alternative plans can be evaluated to assist planners and decision makers in selecting the appropriate plan. At the implementation stage, a GIS can be used to evaluate development proposals, development control, zoning, approval of subdivision schemes and impact studies of large-scale development projects. Finally, at the monitoring stage, it can help to detect land-use and environmental changes, and to monitor whether or not urban and regional development is progressing according to the plan.
>
> YEH 1991

CHAPTER 2
Components of a GIS

A geographic information system consists of data, software, hardware and people (Fig. 2.1). The GIS data is regarded as the most expensive part of a GIS. GIS software is capable of integrating a variety of data types from various sources and providing multiple data entry options. The computer is the most important part of the hardware together with some peripherals, such as digitizers/scanners, high-resolution monitors and plotters. The people tie the data, software and hardware together to perform various functions of data collection, management, analysis and presentation.

2.1 The data

Most geographic information systems are not a part of 'turnkey' technology (providing for a

Fig. 2.1 *Component of a GIS*

supply of equipment in a state ready for operation). When you have purchased a GIS system and the equipment, you then need to acquire the data to run the system. If a GIS is a vehicle, the database management system is the engine; and the *data* or *database* is the fuel of a GIS. The power of a GIS depends upon the quality and quantity of the data used and the performance of the database management system.

2.1.1 Database modeling

Database modeling allows a database to be used through a combination of hardware and software facilities and operations. A GIS should include 'integrated data base management software' (Smith 1987), which would need to be designed to provide:

- the system ability to support multiple users and multiple databases
- efficient data storage, retrieval and update of the data
- nonredundancy of data
- data independence, security and integrity

In the GIS community, geographic information traditionally is divided into two generic classes: spatial data and attribute data. The spatial data should be represented by geometric entities. Although the design of a spatial database and attribute data are interrelated, there

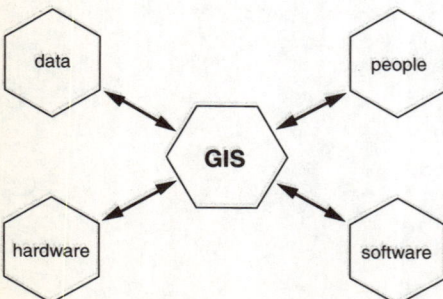

are some fundamental differences between the *spatial data modeling* and *attribute database design*. Present database management system (DBMS) technology is very good for managing attribute data in the form of record and field notes, but mapping involves much more than just the storage and retrieval of attribute data. DBMS technology is not very effective for updating and managing spatial data. When, for example, a land parcel's boundary is adjusted, the surrounding parcels must also be adjusted, and the geometric attributes of all these parcels then change. Manipulating spatial data is much more complex than managing tabular data.

The most popular *GIS database modeling* is to design a georelated database, which can handle two classes of data: spatial data and attributes. First of all, it classifies geographic information into a series of independently defined layers or coverages (Fig. 2.2), each representing a selected set of closely associated geographic features, such as cadastral boundaries, roads and land use. The layer defines the spatial data and attributes in a relational database management model.

The spatial data is classified as areas, lines and points. Each could be defined as either simple cartographic data or data with topological relationship. Spatial data should be distinguished from the data for cartography or digital mapping data. It could be basically described as no-color, no-width and no-shape. The only information that it contains is the geometric and topological characteristics of geographic information.

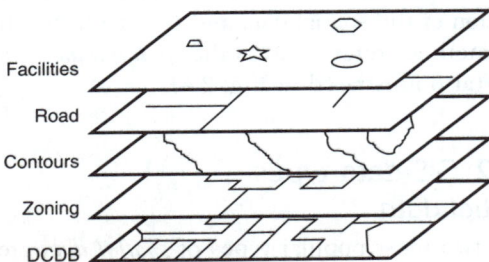

Georelational geographic information database management system is a hybrid DBMS. In the history of GIS development, a number of database structures have been used, such as flat file, hierarchial file and networks structure; these were eventually replaced with *relational databases*. Now, most databases in most organizations are relational databases.

Relational database technology

Relational database technology allows the GIS to manage tabular files in a relational manner. It consists of a series of files named *relations* or *tables*.

A *table* is defined as a set of data elements in a matrix of rows and columns in a relational database system (Huxhold 1991). A table normally has a limited or specified number of columns, but the number of rows it can have is limited only by hardware, the operational system and software itself. Two or more tables are related by the common elements, and can be joined to form a new table (Fig 2.3).

Managing a database in this way could lead to the reduction of duplicated data, and it has unlimited flexibility in forming relationships among the elements in the database.

CASE tools

Application system development is now much too complex to be attempted without disciplined,

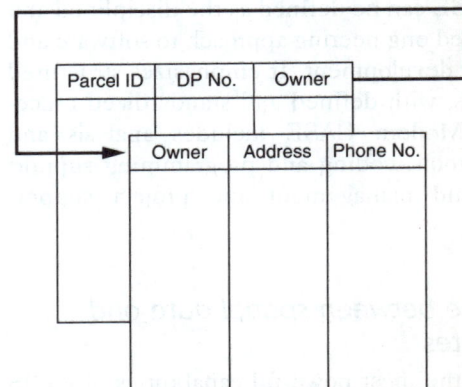

Fig. 2.2 *Layers in a GIS*

Fig. 2.3 *Relational database*

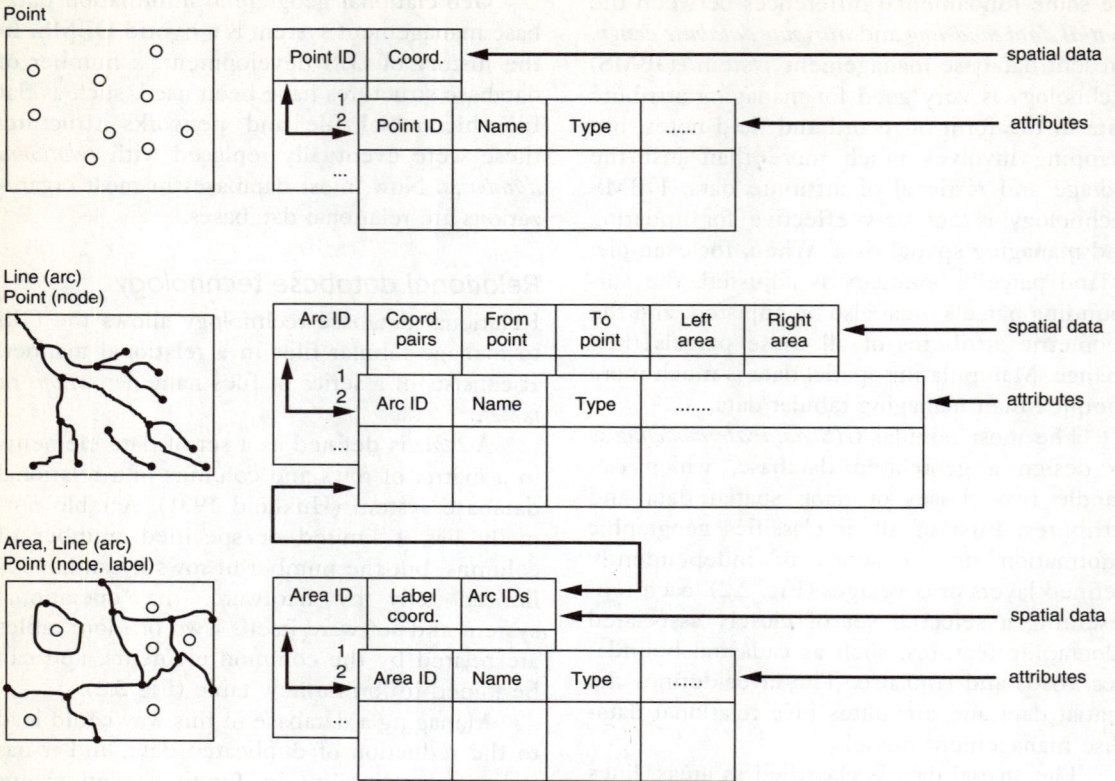

Fig. 2.4 *Linkage between spatial data and attribute*

standardized methods for managing it; hence, *Computer-Aided Systems Engineering* (CASE) has emerged.

CASE can be defined as the disciplined and structured engineering approach to software and systems development. It emphasizes structured methods, with defined and standardized procedures. Modern CASE includes analysis and design tools, coding and programming support tools, and management and project support tools.

Linkage between spatial data and attributes

One of the most powerful capabilities of a GIS is the integration of spatial data and attribute data. The linkage between spatial data and

attributes is through a feature number, whether for:

- point identification
- line (arc)/point (node, label)
- area, line (arc)/point (node, label) (as shown in Fig. 2.4)

The feature number is associated with the location of the spatial data and the attributes. It is sometimes referred to as the *spatial data identifier* (also illustrated in Fig. 2.4).

2.1.2 GIS data types
Spatial data

The two most popular types of *spatial data* are raster and vector. The basic difference between them is that raster records spatial data using cells, while vector data records spatial elements as sets of coordinates (Fig. 2.5).

Raster data

In a raster format, physical features are divided into a rectangular (normally square) grid. Then a matrix of cells is created. Each grid has coordinates, which simply relate to the number of cells away from the original, in both x and y directions. Cells could be assigned values, which together on a map represent geographic features, that is land use, slope, soil type, and so on. Using the same coordinate system, raster data sets can be logically organized into layers or themes for geographic data management and analysis.

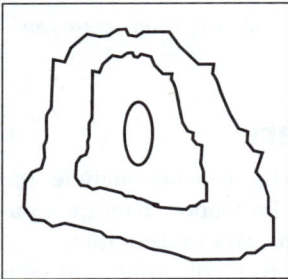

Analog

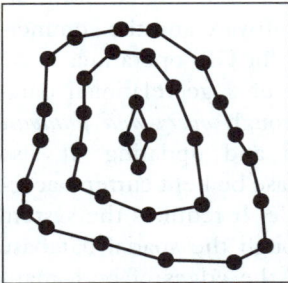

Vector

Raster

Fig. 2.5 *Spatial data types*

Although most (advanced) geographic information systems are based on vector structures, there are also good arguments in favour of *raster structures* for a GIS. Many operations (for example, overlays, spatial statistics and model integration) are simpler and faster in the raster than in the vector domain. Consequently, input of raster data in a GIS is essential for many systems. Possible sources for raster input include:

- classified remotely sensed data
- rasterized versions of vector (cartographic) data
- interpolated point or profile measurements
- scanned maps and photographs

Vector data structure

Vector data is stored as geographic elements that have positional coordinates. But *topological vector data* structures provide additional intelligence. They tell the computer which cartographic objects are connected logically to each other; that is, a polygon is bounded by certain lines that have certain beginning points and end points. Thus topological data structures define how points, lines and polygons are related to each other on a map—an implicit relationship that is usually obvious to the human eye, but not explicitly defined to a computer when it reads through the cartographic records as they have been presented so far.

3-D surface representation

Unlike two-dimensional maps, surfaces can be considered as a special class of spatial objects, where the surface is defined by the data (Gatrell 1990).

A variety of representations are possible, such as:

- contours (the traditional cartographic approach)
- Digital Terrain Model (DTM)
- a grid of elevations (a raster)
- a Triangulated Irregular Network (TIN), where triangular facets are used to represent the surface

Coordinate Geometry (COGO)

Bearings and distances from a known point are used to define the locations of other points, and are commonly found in cadastral applications where the surveyor's field observations can be directly inputed into the GIS. Most systems that use this approach also compute coordinates for all points, and store these as the location of each point. The bearing and offset from one feature can be used to define the location of another; for example, utilities (gas, water and electricity) are usually a standard offset from the property boundary.

The conversion between raster and vector data

Raster and vector data type can be converted through *rasterization* and *vectorization*. Rasterization is carried out through griding a vector data set; vectorization polygons from a grid is conducted by following the boundaries between different cell values. However, vectorizing linear features from a grid is more complex and requires more sophisticated operations (Fig 2.6).

Attribute data

The types of attribute data is relatively simple compared with spatial data. Each GIS system may have its own definition, but generally, it could be classified as types of character, numerical data, logical data, date, memo and image file (RS image, scanned photograph) and so on. Spatial data types could be used to illustrate the attribute, such as image, graphics, and so on (Table 2.1).

Table 2.1 *Attribute and data type*

Attribute title	Data type
Name of an area	Characters
Land suitability	Characters
Overview of an area	Characters
Total population	Numerical (Integer)
Average income etc.	Numerical (Floating)
Date of last election	Date
Outlook of property	Image

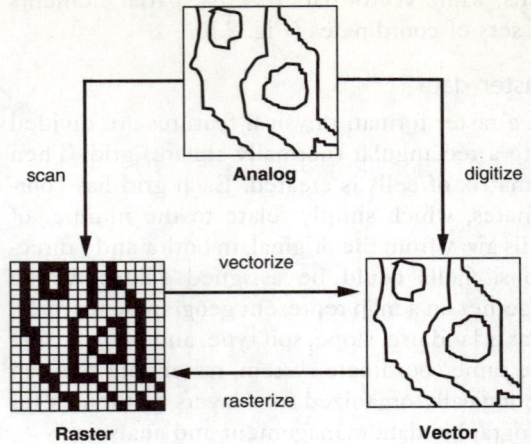

Fig. 2.6 *Conversion between data in raster and vector*

2.2 The software

The capabilities of GIS software include the data collection and maintenance, data manipulation, data analysis and data presentation.

Data collection involves collecting and converting georelated data into digital form as specified by the GIS software and the commercially available DBMS for GIS operation.

The management of a georelational database is performed through *query and database* structure modification and updating. It also requires that the database be kept current, accurate, integrated and safe. It requires the system to run seamlessly through the spatial database and attributes; that is, the edges of both maps will, or can, be joined in such a way that there is continuity of all lines across the edge.

Analysis involves the integrated operation among several layers or parts of spatial information, extraction of intelligent information from the database, and the creation of new georelational information, new spatial features and its georelational spatial database.

Data presentation includes display on the screen and digital mapping on various materials, and the provision of digital data in a particular format.

Most GIS adopt *Graphic User Interface* (GUI), which defines how a user interacts with a computer. GUI provides an easy-to-learn, easy-to-understand and easy-to-operate environment for running a GIS.

The user interface exploits the benefits from user interface standards and very high performance computers, such as workstations and high-end PCs. The user can manipulate GIS through a mouse and/or a keyboard.

Standard GUI makes full use of the range of concepts such as canvases, panels, list-views and alert-boxes. It also exploits many graphical devices, such as pop-up menus, button-items, sliders, choice-items, toggle-items, text-items and label-items.

Another very important feature of GUI is the native language support function within the same user interface design. All input and output messages are provided in English, as well as in other languages.

2.2.1 GIS data collection

Geographic data for input to a GIS is acquired in a variety of formats, including graphic (spatial) data and nonspatial (attribute) information. This comes from both printed data and digital files, and from digital spatial data tapes, such as remotely sensed (satellite) images or digital elevation data. Often these data will require manual or automated preprocessing prior to encoding. During data acquisition relevant information for each data type should be obtained, which as far as possible describes the accuracy, precision, currency and spatial characteristics of the data that can be stored in a metafile (Smith 1987).

Format conversion

Format conversion includes digitizing and data import. Digitizing converts the spatial features on a map into digital format, while data import converts existing digital data, either spatial data or attribute data, into the formats that can be part of your geographic database.

The two conventional methods of digitizing are with a tabular digitizer or a scanner. A GIS normally has digitizing functions or modules that link the computer and digitizer/scanner, along with recording the geometric data in vector or raster format. Some geographic information systems also have the function that can convert data between vector and raster format.

Digitizing with a tabular digitizer converts points, lines and area features that compose a map into coordinates. A single coordinate may represent a point feature, and a string of coordinates may represent a line, or one or more lines, that constitute an area.

Scanning technology and *raster-to-vector conversion algorithms* provide new opportunities for automatic image to line coverage. Scanning has proven to be more efficient than manual digitizing, especially for documents with a considerable amount of linear features.

Vectorization could be provided with either on-screen digitizing, or automatic or semiautomatic features extraction.

Some software can carry out digitizing across multiple scanned map sheets, eliminating delays caused by edgematching independent map sheets. The digitizing may be carried out by multiple users into one seamless mapbase, thereby eliminating delays in merging many independent datasets.

By tracing raster images, the features can be extracted automatically or semiautomatically. Some software can work with images of any size of resolution, and use a combination of user-defined colour filters and parameter setting to tailor the *tracing algorithms* for production-oriented raster-to-vector conversion.

Data import could be from many sources, such as raster remotely sensed data, global positioning system (GPS), soft photogrammetry* and total station recorded data in ASCII (American Standard Code for Information Interchange) text files, other standard data exchange files, and data generated in other systems. The GIS software should have modules to convert *multiple data structures* into useful structures.

---

* Soft photogrammetry is the digital scanning of air photographs, applying normal analytical photogrammetric techniques.

Format conversion for spatial data can be summarized as shown in Figure 2.7.

Construction and error detection

For vector-based GIS software, the *construction of topological information* is an important function. Topological relationship, such as the information referring to the right and left polygon of a line, arcs that constitute an area and ID for an area, will be automatically created.

Edge-matching

Edge-matching is the process of joining lines and polygons across map borders and boundaries in the creation of a seamless database. The join should be geometric as well as topological; a polygon so joined should become a single polygon in the database, and a line so jointed should become a single line segment (Fig. 2.8).

Adjustment for data integration

Scale change is the ability to perform the operations associated with change of scale, which may include line thinning and generalization. When integrating map tiles or source maps with different scales in digital form, it is necessary to change some or all of the map scales to a unique scale for data integration. In such operations it may be required to reduce the number of points

of an arc when the scale is reduced, and some of the smaller features may reduce their dimension, for example, from an area to a line, a point and so on.

Rubber sheet stretching is the ability to stretch one map image to fit over another, given common points of known locations. It is widely used in the edgematching process when two known points at separate map tiles cannot be linked.

Coordinates/Projection change is the ability to transform maps from one set of map coordinates/projection to another. Since the earth is a sphere, source maps usually represent real-world coordinates that have been projected onto a flat surface. There are hundreds of projections available in the world. All projections generate different degrees of distortion. Some of the most widely used projections are as follows:

Orthographic

The perspective of the Orthographic projection views the globe from an infinite distance. This perspective gives the illusion of a three-dimensional block. Distortion in size and area near the 'projection limit' appears more realistic than in most other methods of projection.

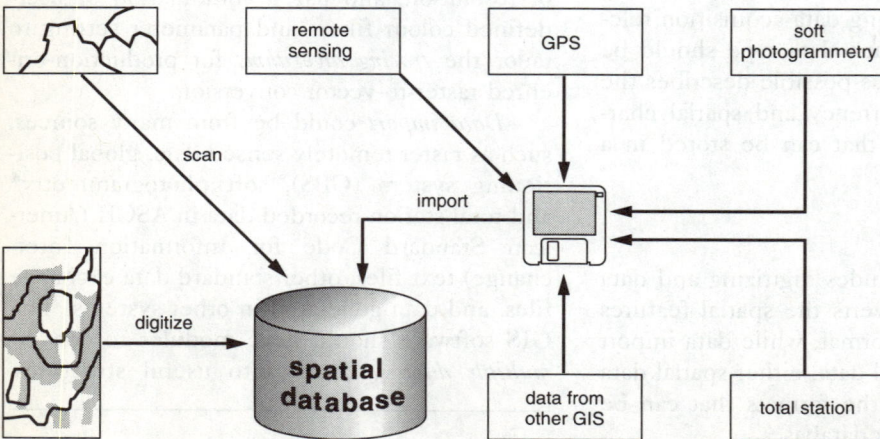

Fig. 2.7 *GIS data acquisition*

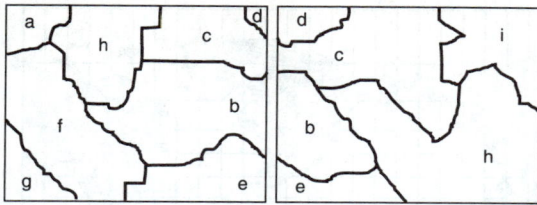

Before edge-matching

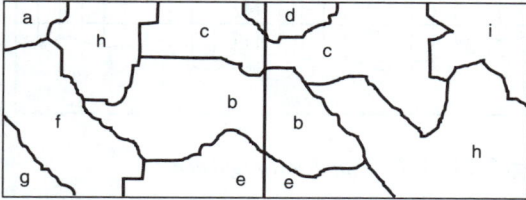

During edge-matching

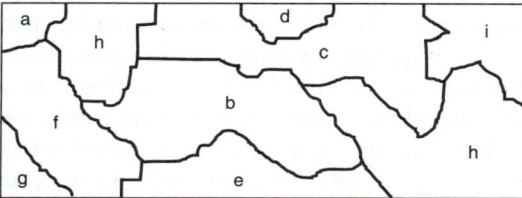

After edge-matching

Fig. 2.8 *Edge-matching processing*

Robinson

In this pseudocylindrical projection, meridians are equally spaced and resemble elliptical arcs, which are concave toward the central meridian. Refer to Figure 2.9. The central meridian is a straight line 0.51 times the length of the equator. Parallels are equally spaced straight lines between 38° north and south; spacing decreases beyond these limits. The projection is based upon tabular coordinates instead of mathematical formulas.

Mercator

The Mercator projection (Fig. 2.10) was created originally to display accurate compass bearings for those travelling on the seas. An additional feature of this projection is that all local shapes are accurate and clearly defined.

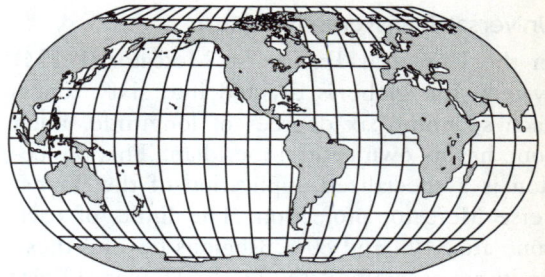

Fig. 2.9 *Robinson projection*

Transverse Mercator

The Transverse Mercator projection is similar to the Mercator except that the cylinder is longitudinal along a meridian instead of along the equator. The result is a conformal projection that does not maintain true directions. The central meridian is centered on the region to be highlighted. This centering on a specific region minimizes distortion of all properties in that region. As meridians run north and south, this projection is best suited for land masses that also stretch north to south.

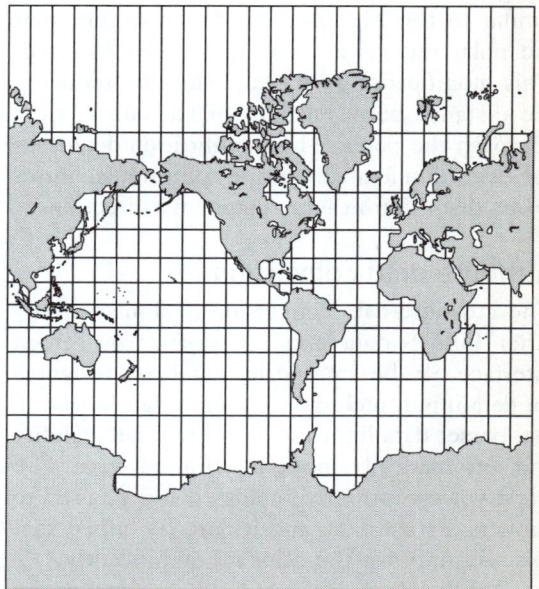

Fig. 2.10 *Mercator projection*

Universal Transverse Mercator

In the Universal Transverse Mercator (UTM) system, the globe is divided into sixty zones, each spanning six degrees of longitude. Each zone has its own central meridian. This projection is a specialized application of the Transverse Mercator projection. The limits of each zone are 84°N and 80°S. Regions beyond these limits are accommodated by the Universal Polar Stereographic projection.

Alberts Conic Equal-Area

The Alberts Conic Equal-Area projection uses two standard parallels to reduce some of the distortion produced when only one standard parallel is used. Although neither shape nor linear scale is truly correct, the distortion of these properties is minimized in the region between the standard parallels. This projection is best suited for land masses that extend more in the east-to-west orientation than those lying north to south.

Miller Cylindrical

The Miller Cylindrical projection (Fig. 2.11) is similar to the Mercator projection except that the polar regions are not as aerially distorted. This modification is accomplished by reducing the distance between lines of latitude as they approach the poles. The modification decreases the distortion in area, but the compromise introduces distortion in local shape and direction.

Attribute data conversion

The attribute data conversion is similar to any other DBMS data entry. User-friendly screen interface can be created by database designers for data input, including frequent data selection list, preset data limit to control the data validity and automatically input interlinked data. The latest voice-control technology could be used to convert textual data, and it is to be hoped that this will improve the accuracy significantly.

Another aspect of attribute data conversion is using existing digital data in various format, from simple ASCII data to spreadsheet data to

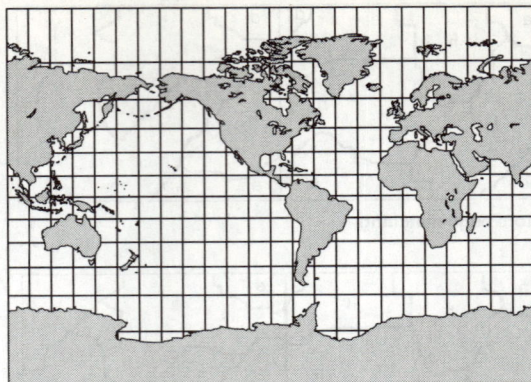

Fig. 2.11 *Miller cylindrical projection*

database data. Before the data is of any use, the data has to be converted into the current database format and some adjustments made, such as units of quantitative data and items of a classification. A column should be defined as a common column to link with existing databases.

For dynamic data exchange, which may happen when a GIS needs to access a corporate database, an interface would need to be established to convert data automatically.

2.2.2 Database manipulation

The database management functions help to organize information into a meaningful structure by providing viewing, editing, manipulating and querying. Through database management, a variety of individual information needs can be satistified, and the value of a georelational database can be kept.

Viewing database

The function of viewing a database provides a handy tool for users to obtain a glimpse of the georelational database, while performing various query and analysis tasks. A GIS normally has the following functions:

- *Browsing* is setting up a monitor and browsing through the spatial data, attribute data and files associated with various graphic information.

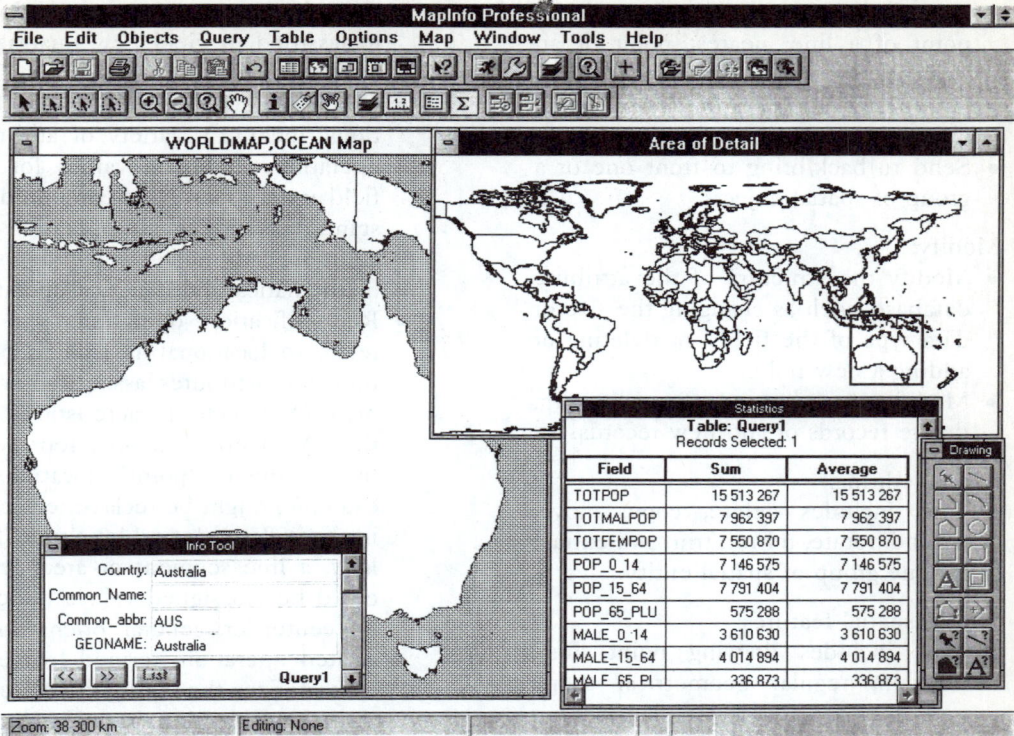

Fig. 2.12 *An example of friendly interface with menu, icon, multiwindows and zooming tools*   MAPINFO

- *Panning* is viewing part of the spatial data without changing the scale, and viewing part of the attribute database, using scrolling key/bar.
- *Zooming* is viewing the graphic data with a different scale to enlarge or reduce.
- *Multiwindow* is viewing different parts of the geographic information files or different geographic data files, in different windows and different scales on one or more monitors.

Editing database

The editing function of a GIS allows the user to modify the structure and contents of a geo-relational database in order to update the contents of the database, regulate data imported from or exported to other databases, correct mistakes found, and optimize the structure of the

database, and so on. Versatile *select tools* are very useful when dealing with multiitem/object editing. When there are a large quantity of repeat editing tasks, *macros* for certain procedures could be used to improve the efficiency of editing:

- Select:
    - Select a particular attribute file.
    - Select/unselect particular spatial data or a group of spatial data with various devices, keyboard, mouse and/or digitizer, and so on.
    - Select spatial data on point within/across a rectangle or circle window, and generate irregular geometric polygons for selecting information within the area.
    - Provide other tools including snap and gridline—on/off, and set distance in

two directions; end point, middle point of a line; nearest point of an object, and center of areas; and so on.
- Group/ungroup two or more spatial objects.
- Send to back/bring to front one or a group of spatial objects.

- Modify:
  - Modify the structure of the attribute database, such as changing the size or data type of the fields, or deleting or adding a new field.
  - Modify the content of each item, delete records or add new records.

- Spatial editing:
  - This includes delete, copy, move, extend, rotate, mirror, trim and extract for any group of spatial entity.

- Create spatial features:
  - This includes drawing point, line, polygon, regular geometrical shape and other defined spatial shapes.

Manipulating database

Manipulating function in a GIS is to process an integrated georelational database for further operation. To improve the efficiency of query, analysis and presentation, databases are required to be organized in a particular order, containing appropriate and comprehensible information.

- Sort & index:
  - Sort and index the georelational database according to particular criteria.

- Join & extract:
  - Join full or part of two or more geo-relational databases.
  - Extract part of the georelated database from one or more georelated databases.

- Reclassify (Berry 1987):
  - This involves ranking or weighting qualitative attributes to generate a new attribute with quantitative values.

- Quantitative attributes may also be reclassified to yield new quantitative attributes.
- Other 'quantitative reclassifying functions' include a variety of arithmetic operations involving values for each field and a specified or computed constant. Among these operations are addition, subtraction and other scalar mathematical and statistical operations.
- Reclassification operations can also relate to locational as well as purely thematic attributes associated with a map. One such characteristic is position. An overlay category represented by a single 'point' location, for example, might be reclassified according to its latitude and longitude. Similarly, a line segment or areal feature could be reassigned values indicating its center or general orientation. A related operation, termed 'parceling', characterizes the contiguity of categories. This procedure identifies individual 'clumps' of one or more points having the same numerical value and which are spatially contiguous.
- Another locational characteristic is size. In the case of map categories associated with linear features or point locations, overall length or number of points might be used as the basis for reclassifying those categories, such as height or altitude.
- In addition to the value, position, contiguity and size of features, categories may be reclassified on the basis of shape. Categories represented by point locations have measurable 'shapes' in so far as the set of points implies linear or areal forms. Characteristics of shape associated with linear forms identify the patterns formed by multiple line segments. The primary characteristics of shape associated with areal forms include topological genus, convexity of boundaries and nature of an edge.

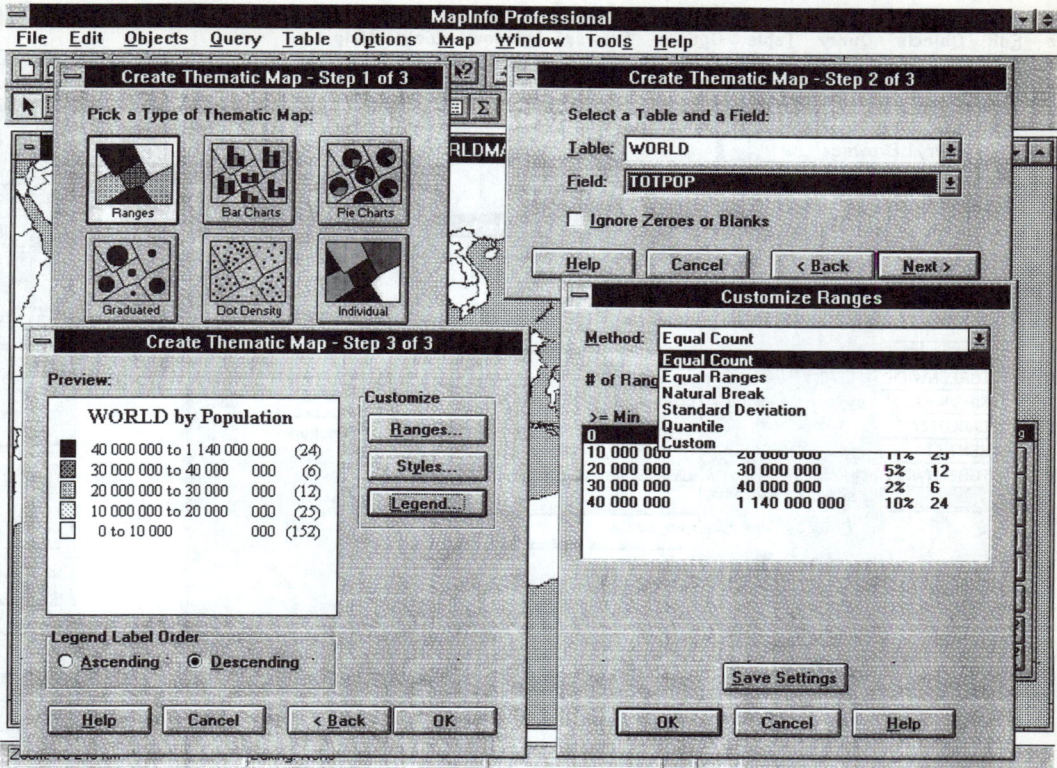

Fig. 2.13 *Classification function used in thematic mapping* MAPINFO

Querying database

Query is the preliminary use of a georelational database to generate useful information to support the information needs at various stages of planning and decision making, management and operation.

- Structured Query Language (SQL) is nonprocedural query language used to access relational databases, having an English-like syntax and allowing for data insert, update, query and protection (Van der Lans 1988).
- Basic spatial query
  - *Points:* The two typical measurement activities associated with points are enumeration of a total number of points, and enumeration of a total number of points falling within polygons or on lines.

- *Lines:* Activities associated with lines are point to point measurement and measurement along a curve line either contained in or bordering an area.
- *Areas:* Activities associated with areas are measurement of the area of a polygon and the perimeter of a polygon area.
- *Volume:* Volumetric measurement is performed either through a cross-section technique or through overlays of multiple surface.

- Tell me:
  - Select geographic features from spatial data, find the attribute related to them, and display the information in tabular/record form, together with the map.

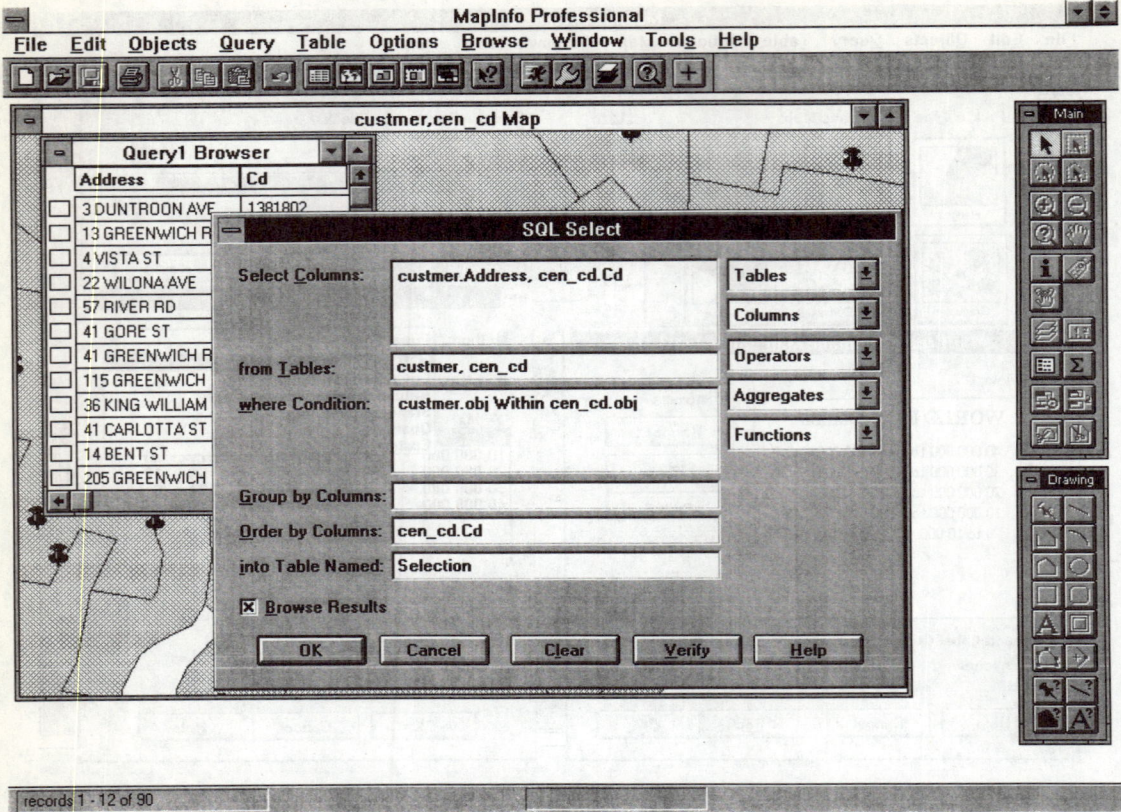

Fig. 2.14 *An example of SQL function operating between different databases* MAPINFO

- Show me:
  - Specify the value or the range of the value, and logical and algorithmic operation over particular fields from the attribute database, and specific search limit on the spatial database, to find the geographic features on a spatial database and display on the screen, such as map land parcels within a radius of 30 kilometres (18.6 miles) of the Central Business District (CBD), with expansive (clay-type) soils, poor internal drainage or other characteristics that make them unsuitable for housing development (AISIST 1993b).

- Other query:
  - *Trends:* What has changed since ... ?
  - *Patterns:* What geographical patterns exist?
  - *Simulation:* what if ... ?

2.2.3 Spatial analysis and interpreting capabilities of a GIS

The crucial factor that distinguishes spatial analysis from traditional information systems is the use of locations for referencing information as an important variable in quantitative analysis. Spatial analysis is not an independent technology of a GIS. It must be used together with database management and digital mapping, along with image processing and statistical analysis. By exploiting the spatial dimension, spatial analysis introduces a new perspective that can greatly enhance decision making and problem solving.

The two basic approaches in representing locational information are vector, based on point/line/polygon, and raster, based on grid. The differences between them are significant in terms of strategies for implementation and other issues, but basic analytical functions are the same. The spatial analysis functions of a GIS will be illustrated with a vector-based GIS.

Most GIS software vendors, however, distribute their analytical and interpreting capabilities of a GIS in the form of modules. These modules could either be integrated with core GIS software or run independently.

Buffer generation

Buffer generation involves the creation of new polygons from points, lines and polygon features within the data bank. Buffer could also be generated around a single point, line and polygon or around a series of such features.

It is a very useful tool to analyze the impact of certain spatial features and events on the others. Examples include a study of the area that has appropriate access to river water, the impact of urbanization on rural and semirural communities, and declaring the affected area in the case of poison leaking.

Nearest neighbor search

Nearest neighbor search is the ability to identify points, lines or polygons that are nearest to points, lines or polygons specified by location or attribute.

Connectivity and contiguity analysis

The ability to identify areas, lines or points that are connected to other areas, lines or points, and to identify areas that have a common boundary or node is normally called *contiguity analysis*. The integration of the nearest neighbor search, and connectivity and contiguity analysis is called *proximity analysis*.

Polygon overlaying operations

Most geographic information systems store data in layers, which are classified by point, line and polygon. It makes little sense to overlay in between different spatial entities to create a new heterogeneous data layer, such as point and line. There is little opportunity to do point to point and line to line overlay, except to aggregate several similar themes, such as integrate schools or hospitals into a facility theme, or join the road system with bushwalking tracks, and so on.

Overlaying polygon with polygon is widely used in spatial analysis. The overlay operation may or may not involve the creation of a new polygon. For example, the overlay of national, state and local government administrative boundaries does not create new polygons because they are homogeneous in attributes. The overlay of zoning information and cadastre boundaries does not need to create new polygons while the zoning information uses the cadastral boundaries; it is unlikely that one piece of land will be classified into two different zones.

When the complex physical world is generalized, classified and abstracted for each kind of characteristic by human beings, the collection of the location with identical attributes forms a recognizable area. For the same area, different attributes may have their own classification and retain class boundaries. When two or more attributes are considered in a spatial analysis, further classification may be required through polygon overlay operation. A new data set is created, containing new polygons created from the intersection of the boundaries of all separate polygon layers.

The reverse operation is *map dissolve*, which has the ability to extract the single attribute from a *multiple attribute polygon* file.

Network analysis

Network is a system of interconnected linear features through which the mass or the information is transported. In the physical world, *network* is used for transport, distribution and communication. Typical networks are:

- roads, railways, waterways and air corridors, and even the transport of people and the distribution of goods and services
- utilities—transport and the distribution of gas, electricity, water supply, sewerage, storm water, and so on
- communications—telephone, computer network, cable TV wireless communications, and so on
- geocoding—the address of a particular object on the road system

There are several typical operations in network analysis, such as the following:

Routing

When more than one terminal is involved, one needs to plan a routing to set the order and select the path.

Tracing

Sometimes in a network, the connectivity of network has to be determined. For example, who would be affected if the electricity supply for a particular area was to be shut down for a maintenance reason? Or, in the case of finding pollutants in the river, one needs to know the possible sources of the pollutants by reference to a map showing all tributaries upstream.

Allocation

It is a traditional *operational research* problem to play with the demand, supply and impedance. It is an operation to assign area to a location based on predetermined criteria and network operation. For example, a bank staff member may want to know the limits of accessibility within 10 minutes drive of the bank.

Real-time monitoring

Equipped with GPS and a communication device, a GIS can now show the position and time for dynamic objects in a transport network or in a study area.

Quantifying surface analysis

Surface analysis is a digital terrain analysis through a representation of terrain relief in a computer readout format. Its operations include the following:

- *Interpolate elevations and specific interval contouring:* This is the ability to take random or regularly spaced data and generate a grid or other structured model framework on which automated contouring can occur. Based on this interpolated model, contouring can be drawn out.
- *Surface slope aspect:* Calculate the slope aspect.
- *Area and length calculation:* Calculate the surface area and path along a surface.
- *Generation of cross-section and long section:* Generalize a section view at a particular location
- *Volumetric cut and fill calculations:* Create a pit or peak and calculate the volume.
- *3-D representation:* Create 3-D display wire-framed surface and shaded relief.

Image processing

Preprocessing

Radiometric corrections

The radiance of each pixel is measured and classified. Atmospheric scattering and absorption affect the image quality. Suitable adjustments can be made to correct the distortions.

Geometric corrections

Geometric corrections are made to remove the errors caused by slight wobbles of the satellite, change in the scan mirror position, platform velocity and the earth's rotation. The corrections are done on the basis of some prior knowledge of terrain.

Restoration

The *restoration* process is specifically aimed at the removal of faults in the image, such as 'banding' effects.

Image enhancement

Image enhancement is done to facilitate visual interpretation so that certain kinds of information stand out.

Contrast stretching

As the atmosphere, the sensor capacity and the features of the scene itself may reduce the contrast in an image, the brightness value of an image can be uniformly or selectively stretched.

Spatial filtering

Spatial filtering enables the improvement of images through the enhancement of spatial features, directions or textures.

Random noise elimination

All images are subject to some degree of noise. One of the easiest methods of adjusting for noise is to take an average of the count values of each pixel and its nearest neighbours.

Classification and analysis

Principal component analysis

Multiband images may show a certain degree of similarity between spectral bands. Most image-processing systems have the ability to display three bands through the use of the three primary colours. This is not very useful for those images that contain more than three bands. This is especially evident if bands overlap, or where a physical correlation between features on the image is evident.

As a means of solving this problem, the image is typically transformed through the introduction of new bands, which are linear combinations of bands already in existence in the base image. The principal component analysis is used to remove the correlation and to maximize the separation between bands.

Density slicing

The technique of *density slicing* involves the grouping of regions with similar count values and the subsequent representation of these groups in one uniform colour.

Supervised and unsupervised classification

In *supervised classification*, a sample area is selected and its spectral signatures are analyzed, based on field surveys and photo interpretation of aerial photos. Following ground check and sampling, different colours can also be assigned to each spectral data.

Unsupervised classification is used when a ground check of a study area has not been done. The computer is required to distinguish different features on the basis of clustering pixels of similar gray tones or reflectance values.

2.2.4 Data presentation (digital mapping)

GIS mapping products are not limited to maps, but include the tables, graphics and textual documents, or the combination of them. The products media could be not only the hardcopy output of digital geographic information, but also the temporary display on the screen and overhead projector, and a final copy on the film, even in digital form on magnetic tape, CD, and so on. Whatever the content and output media are, GIS products are focused always on the mapping, and GIS software normally provides sound cartographic tools for mapping needs, including cartographic design, display and production capabilities. The 'interactive display functions' normally provide a good preview for the hardcopy of the final products. There is usually a preference to use digital mapping instead of data representation. The table, graphics and documents are considered as auxiliary contents of a digital map.

The digital mapping of a GIS generally consists of the following general procedures:

The objective of the map

Mapping in a GIS could be simply the representation of the GIS database with cartographic elements. It could be the graphic representation of data query and data analysis, a theme, general purpose maps such as topographic thematic maps and planimetric maps, and an even more complicated integration of maps, graphics and documents.

The first step is to make decisions about the layout of the information to be included. This covers not only the geographic information (spatial data and attribute), but also the units, projection, area, titles, legends, scale and a wide range of marginal information.

In creating a map we are in effect performing a transformation of some part of a geographic information into an abstract representation, using some form of symbols. The transformation from geographic information to map relies upon several processes. The two main processes are usually referred to as symbolization and generalization.

Generalization

Generalization is the reduction in the amount of detail that can be shown as the scale is reduced. As the scale becomes smaller, so the represented area decreases in detail; therefore, the number of features shown must also decrease if the map is to remain legible.

The main processes involved in generalization are selection, classification, simplification and combination. Exaggeration and displacement may also be required as the scale is reduced.

Selection processes include selecting spatial data and attributes from the spatial and attribute database. This will depend primarily on the purpose of the map, but the number of categories that can be included will decrease with the scale.

Classification refers to the grouping of attribute data or its ordering or scaling. In the simple cases this may refer to the grouping of statistical data into a small number of discrete classes, based upon some class interval scheme. It also refers to the classification of nonnumerical information, such as vegetation or soil types.

Simplification is the determination of the important characteristics of a feature, the elimination of unwanted detail, and the retention and possible exaggeration of the important

aspects. Simplification refers mainly to lines—linear features and the boundaries of areas—but may also apply to points when symbolized by pictographic symbols.

Combination is a difficult element of generalization to quantify. A large-scale map may show all areas of woodland. At a smaller scale, one could simply select the larger or more important woodland areas, but a truer representation may result from the combination of several small areas into one larger area, often with a simplified boundary and class.

Symbolization

Symbolization is the use of the graphic variables for the representation of geographic features on a map. The design of symbols thus involves the use of point, line, area and text, with variations of type, dimension and color. Color itself has the three variables of hue, lightness and saturation. Some refer to the graphic variables as shape, size and color. It is summarized in Table 2.2.

A GIS normally maintains a library to store the type, pattern of point, line and area (hatching and shading), and so on. Depending on the capability of hardware (plotter and monitor), a GIS also supports a number of colors. A GIS can also specifying the globe type, size and other parameters of symbol, line, shade/hatching for area and text.

Annotation

Annotation is text stored as a layer of the feature(s). It is used to illustrate the feature, such as district name or distance between two

Table 2.2 *Variables for spatial features symbolization*

Variables	Point	Line	Area	Text
Type	✓	✓	✓	✓
Pattern	✓	✓	✓	✓
Size	✓	✓	✓	✓
Color	✓	✓	✓	✓
Angle	✓		✓	✓
Justify	✓	✓		✓
Font				✓

places. The text features is often precisely sized, symbolized and positioned in relation to the feature it represents and other features, to produce legible displays and plots.

Creating cartographic elements

A GIS has the functions of adding title, subtitle, borders, grids, keys/legends, scale bars, north arrows, agency logos, reference maps, summary tables and even images on a map.

Display and viewing raster data with vector data

One needs to display one or several images in one or several windows, and register them with earth reference coordinates, and display together with vector data.

Development and customization of symbols

Some geographic information systems provide development tools to create point, line and area symbols to meet special needs. The digitizer and scanner, or editing function, can be used to create the graphic representation of the symbols, and the special function of a GIS will transform the graphics into symbols.

Graphics, tables, documents and maps

Some GIS software can produce statistic graphics, such as bar/pie chart or line chart to enhance the information presented on the map. Other items, such as tables and documents, can also be included on the map sheet (Fig. 2.15).

Output

GIS software has the ability to be configured with major kinds of high-resolution monitors, printers, plotters and other output devices.

2.2.5 Development tools

Some GIS software provides a development environment, through software development tools or kits. The tools range from repackaged computer languages, such as BASIC, FORTRAN and C to RDBMS and 4GL, (fourth generation language).

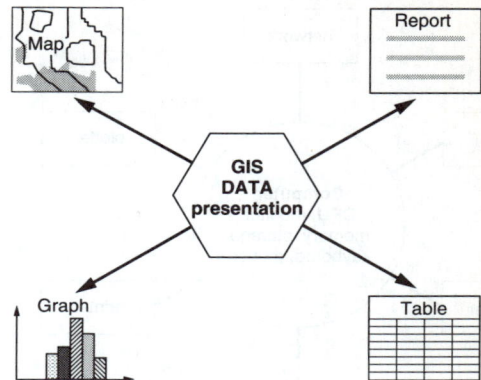

Fig. 2.15 *GIS data presentation*

The development tools can customize and/or produce new interfaces, such as windows, menus, screens and icons, and also create procedures that improve the overall operational efficiency.

2.3 Hardware architecture of a GIS

Hardware for a GIS is basically the same as that for any other information technology. It consists of a computer, storage unit, input/output device and communication network (Fig. 2.16). The hardware setting is close to a computer-aided drafting and design (CAD/CAM) system. But the data capture equipment, such as digitizer and scanner, may show some differences.

2.3.1 Computer

A typical computer consists of:

- a central processing unit (CPU)
- random access memory (RAM)
- secondary storage (hard disk)
- a keyboard and/or mouse
- a monitor

A range of computers are available in the market. At the low end there are *personal computers* (PCs), which are desktop machines using a single-chip processor, usually dedicated to a

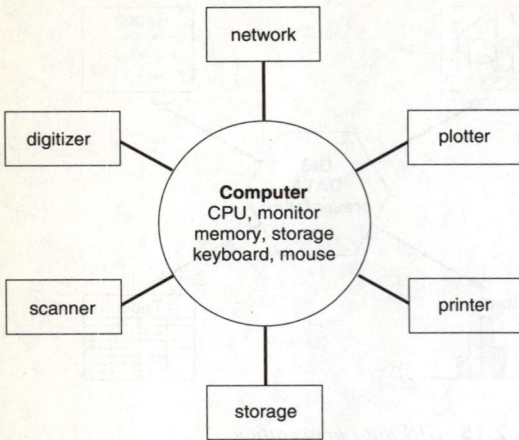

Fig. 2.16 *Components of hardware*

single person. There are stand-alone units or units within a network (Fig. 2.17 on p. 40).

A *workstation* is a computer with both high processing power and large storage capacity. It is almost always based on 32-bit or 64-bit processor, and is generally equipped with tens of megabytes of RAM.

A *mainframe/minicomputer* is normally a central computer in a network. The real difference between a mainframe and an engineering workstation is the input/output capacity, its storage and the applications for which it is used. A typical workstation might have one or two 1-gigabyte disks. A mainframe might have a hundred of them. Workstations are normally used for interactive applications, whereas most mainframes are used for large batch jobs or transaction processing, such as banking or airline reservations where huge data bases are required.

A *supercomputer* is specially designed to maximize the number of FLOPS (floating point operations per second). The bottom line is 1 gigaflop/second. Supercomputers have unique, highly parallel architecture in order to achieve these speeds, and are only effective on highly complex problems.

An *X-Terminal* (Fig. 2.20 on p. 42) is similar to a workstation in that they both have a keyboard, mouse and a high-resolution graphics

screen, but they do not have a local disk drive or a general-purpose computing CPU (LIC 1993). Furthermore, although their specialized centralized processing unit can perform powerful graphics work, they rely on other computers elsewhere in the network to supply the information to be displayed.

Each type of computer can be found in various GIS applications, such as:

- stand-alone desktop mapping system for marketing research
- facilities management running on a workstation
- the combination of a PC and a workstation for an urban GIS
- mainframe/minicomputer used as an 'attribute database management system' in a corporate GIS
- supercomputer used for monitoring global climate change

Workstation and PC vs mainframe and minicomputer

From the advent of commercial computer systems in the late 1960s until the early 1980s, mainframe computers and minicomputers were the core of virtually all data-processing systems. A central computer was the host to a variety of terminals and peripheral devices. All data was stored on disk drives connected directly to this central computer. Thus, the mainframe or minicomputer was required to execute instructions, perform computations, and control data inputs, outputs and storage, as well as direct the flow of the data for all users. A homogeneous computer environment was required—one in which all computer resources were supplied by one principal vendor, and its closely allied manufacturers of peripheral devices. Such installations were not always found to be cost-effective.

For GIS users, one of the principal drawbacks of this type of system is that the computer itself becomes a virtual bottleneck for the massive amounts of graphic data that it has to handle interactively.

Workstation and PC with a greatly enlarged RAM and manipulation system now play significant roles in changing the structure of the GIS industry, and are becoming the dominant platform for a GIS. Mainframes and minicomputers have gradually lost their share in this market.

PC vs workstation

Central processing unit

PCs brought computer processing power to a greatly enlarged user's desktop. The significant development of the 32-bit (some 386 or higher) microprocessor produced a dramatic improvement in computer performance over the older 16-bit microprocessor.

But the power of the processor is the principal distinction between the engineering workstation and the personal computer. The current situation is that the workstation is the dominant CPU for a GIS, because of its significant performance.

Operating system

The majority of PCs use DOS (disk operating system) as their core operating system. Workstations, on the other hand, employ many different operating systems, including the various clones of UNIX. As a result, software written for the PC can run on any machine that is 'PC compatible' regardless of the operating system. Software must be separately compiled and tested for each operating system, along with the architecture of each processor. Some GIS software may work only on some specified workstations.

UNIX offers many advantages over DOS, including multiprocessing (the ability to perform more than one process at a time). This is important to a single GIS user who may want to plot a map, search records, run analysis and conduct other tasks concurrently. Moreover, the UNIX operating system allows a file server to serve the needs of several client workstations simultaneously.

The X-Windows standard, or X-11 as it is sometimes known, is another significant feature of a workstation. A user can access and use several applications at the same time, since those applications can all run under the same X-11 interface on multitasking workstations. *Windows* program for a PC cannot provide the same quality performance as X-Windows for a workstation, though improvements are being made each year.

Configuration

Standard PCs have a certain degree of difficulty to run a GIS. It is possible to upgrade a PC to meet the strict requirement of a GIS, such as increasing the RAM, and using add-on network board, a mouse, super-VGA (Video Graphic Adapter) board, a 'display accelerate board', and so on.

The major difference between PCs and workstations is that workstations come out of the box configured to run large applications, such as a GIS. A typical workstation is supplied with a RISC (Reduced Instruction Set Computing) CPU with tens of megabyte (MB) RAM; hundreds of MB hard disks; mouse; high-resolution color display; graphic acceleration; network card; and all the operating system, windowing and network software installed on disk.

The trend

The PCs are still improving in two major aspects. On the one hand, the PC has passed 100 MIPS (million instructions per second) threshold through Pentium (or 586), and the development of the next generation processors is continuing.

When workstations exceed 300 MIPS (*GIS World* 1993), multi-CPU workstations will be commonplace, true 64-bit workstations will appear, workstations will continue to decrease in price to PC prices at the low end, and software applications will make effective use of X-11 and the increasing power of its RAM.

The introduction of Power PC from IBM and so on and Digital Equipment Corporation's Alpha computer will compete with high-end PCs through running an emulated operating system (Ryan 1994), and close the gap between PC and workstation (Strand 1993).

The applications

Applications for which workstations are most suited include generation of geometrically corrected orthophotos and satellite images from raw data, mosaicing images together to provide coverage of wider areas, integrating image-processing system raster-based data with GIS vector-based data, running multiple applications at the same time (for example GIS, image-processing system and DBMS software), and providing access from common data to networks of workstations. While PCs are good at data capture, viewing, and plotting in terms of lower cost, the data may lose precision when transferring data from a 32-bit PC to a 64-bit workstation.

2.3.2 Input device: digitizer/scanner

The digitizer is the device used for digitizing, which is the process of identifying points on a map and registering them in a digital database or file. It is the most common method for converting existing maps and images into digital form. However, it is also very tedious when converting high-density maps.

Normally, the digitizer consists of a fine wire grid embedded in the surface of a table. The underlying electronic wire grid records the location of the digitizer cursor cross-hairs and transmits it to the computer in the form of coordinates. The digitizer cursor normally has four or more buttons, each of which can be programmed to perform certain kinds of commands and functions, such as recording, cancelling and end pointing.

Some geographic information systems can define a command area, which includes a set of commands accessible by the point device on a digitizing table to assist in the digitizing. The operator can use the point device only to perform the digitizing operation.

Optical *scanners* in some situations can replace digitizing by automatically converting hard-copy maps to digital form. They do this through a 'raster-to-vector' process of *optical scanning*, which is the use of a laser beam to scan the surface of a map and convert it to countless tiny dots. These dots are stored as a 'raster image', a matrix of dots with a selected color mostly in black and white, which has a dot for every possible location on the map. Since this matrix requires a very large amount of storage, a more efficient method is used to store the map information in the computer, by combining all the raster dots that make up a line into a vector, which requires only a beginning point, a length and a bearing for storage purposes. Thus, the optical scanner reads the map, converts it to a raster image, and then combines adjacent dots into vectors.

Source documents containing lines representing linear features and boundaries can be scanned to create a *raster structure data set* with very fine resolution; the raster data set can be vectorized to create a *vector structure data set*. The accuracy of the output dataset depends on the map scale of the source documents and the scanning resolution.

2.3.3 The printer/plotter

The *printer/plotter* is the device used to produce hard copy, instead of interactive applications on a terminal. There are many kinds of printers and plotters available since the first and oldest printer: the impact printer.

A *matrix printer* uses a print head containing between seven and twenty-four electromagnetically activiable needles, which scan across each print line. It can produce a resolution of 300 dots per inch. For example, a simple letter could be represented in a 5×7 dot matrix across the line. The print quality can be increased by two techniques: using more needles and having the circles overlap.

A *laser printer/plotter* uses almost the same technology as a photocopy machine (Tanenbaum, 1990). The heart of the printer is a rotating precision drum. At the start of each page cycle, it is charged up to approximately 1000 volts and coated with a photosensitive material. The modulated light beam from a laser is scanned along the length of the drum, and a rotating octagonal mirror is used to scan the length of the drum to produce a pattern of light and dark spots. The spots where the beam hits lose their electrical charge. After a line of dots has been painted, a reservoir of an electrostatically sensitive powder spreads the toner which is attracted to those dots that are still charged, thus forming a visual image of that line. The toner-coated drum is pressed against the paper, transferring the powder to the paper. The paper is then passed through heated rollers to bind the toner to the paper, permanently fixing the image. Later in its rotation, the drum is discharged and scraped clean of any residual toner. Many of these systems translate the input text to an intermediate language called PostScript, and it is then downloaded to the laser printer. Inside the printer is an interpreter that converts the PostScript to pixels and prints them. PostScript is the de facto standard for laser printers.

An *electrostatic plotter* is like a photocopier. It uses an electrostatic charge to attract colored ink tones to the paper.

A *direct thermal plotter* utilizes heat to transfer colored wax film (in cyan, magenta and yellow) onto paper, in a way similar to color photocopiers and laser printers. The combination of proportions of the three colors can produce nearly all possible color combinations at quite a high resolution.

An *inkjet/bubblejet* printer works on the principle whereby minute computer-controlled jets of ink are squirted onto the paper to form text or graphics output.

A *pen plotter* plots by moving either the pens in one axis and the paper in the other (drum plotter), or the pens in two axes.

The evaluation criteria for the performance of scanners/digitizers and printers/plotters can be, but are not limited to, the following:

- Maximum size: the largest-sized document that can be scanned, digitized, printed and/or plotted is AO size (the largest format size for scanners and plotters), and A3 size for laser printers.

- Maximum resolution: the resolution in number of points (pixels per inch) is 500 for a typical large format scanner, and 400 for electrostatic plotters.
- Geometric accuracy is 0.05% for a typical large format scanner.
- Printer may be color or monochrome.
- Host computer may be PC or workstation (Hewlett Packard, Intergraph, Sun, and so on).
- Price: compare product prices.
- Ease of use (for printer/plotter): pen plotters are not user-friendly.
- Speed (for printer/plotter): pen plotters are usually slow compared with the thermal technique.

2.3.4 Storage
Optical disk

Optical-disk storage provides an option for storage with large capacity, but which is comparatively inexpensive, to meet the needs of the GIS data appetite. The majority of optical disks are CD-ROM (compact disc read-only memory). The technology uses a plastic disc that appears identical to an audio CD, but can store approximately 600 MB of data. The disc itself is composed of a reflective layer encased in plastic. Data is encoded by a series of 'pits' (small depressions in the reflective surface) and 'lands' (areas separating the pits) along a continuous spiral track. When the disc is read, a laser is focused on the track and the quantity of the reflected light received by a photodetector indicates whether the optical head is positioned over pit or land (Cartwright 1993). It is suitable for applications where multiple copies of a single large data set are required.

Other optical storage devices include WORM (write once read many) and magneto-optical drives. WORM drives use a laser to write information to media, but once it is written it cannot be changed or deleted. Magneto-optical drives combine both magnetic and optical technologies to offer high-density storage that can be erased and rewritten, similar to the familiar magnetic hard disk.

There are some limitations on optical-disk storage, which include slow data access times and slow data transfer rates. CD-ROM will not replace hard disk, but it serves a complementary purpose.

Magnetic disk

A *magnetic disk* is a piece of metal platter ranging from about 5 to 10 inches in diameter, to which a magnetizable coating has been applied at the factory, generally on both sides. Information is recorded on a number of concentric circles called 'tracks'. A magnetic disk may have two or more platters.

Magnetic disks are installed on almost all workstations and PCs. It is the major on-line disk-storage device.

Floppy disks

A *floppy disk* (diskette) is a small removable medium that is used with a personal computer. Two sizes are commonly used: 5.25 inch and 3.5 inch. Each of these has a low-density and a high-density version.

Unlike the magnetic disks, where the heads float a few microns above the surface, the floppy disk heads actually touch the diskettes. In order to reduce wear and tear, personal computers retract the heads and stop the rotation when a drive is not reading or writing. There is a delay while the motor gets up to speed. The floppy disk is gradually being replaced by optical disks as the media for GIS software distribution.

Magnetic tape

A *magnetic tape* is analogous to a home video tape, but it records data in digital format. It is still the most prevalent medium for both data backup purposes and data delivery.

2.3.5 Computer configuration and network

Stand-alone model

The arrival of the PC and workstation has introduced a *stand-alone*, single-user model of computer configuration. A system consists of a computer and a limited number of peripherals, running stand-alone GIS software (Fig. 2.17).

Centralized model

In a *centralized model* system architecture, users access data by running programs that execute on the mainframe or minicomputer's CPU. Users typically utilize terminals to transmit and receive data from the central system (Fig. 2.18). Although data is centrally maintained, data centers can support users at remote locations.

Distributed model

A *distributed model* is a computer network composed of networking hardware and software, the computers attached to the network, and the peripheral devices attached to either the computers or directly to the network (Fig. 2.19).

The type of network that connects computing machinery in a single building or site is called a *local area network* (LAN).

Another kind of network is *wide area network* (WAN). WAN is typically two or more LANs attached by long-distance telecommunication technology, such as microwave links, satellite relays, fiber-optic runs or leased telephone lines. Point-to-point networks connect mainframe computers using these links.

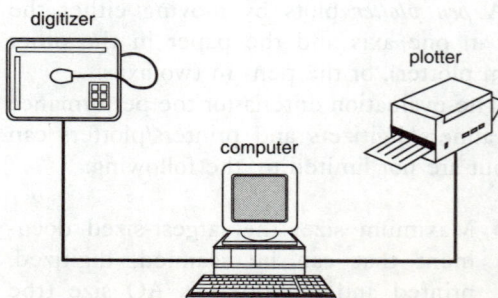

Fig. 2.17 *Stand-alone model*

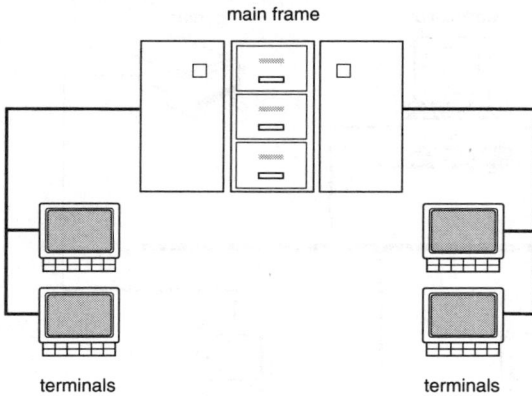

Fig. 2.18 *Centralized model*

GIS technology and system configuration

A GIS, broadly speaking, is applicable technology found in many computing architectures. The stand-alone configuration is limited to certain kinds of applications, such as departmental GIS applications or project-oriented GIS implementation. The centralized model is less in demand in the 1990s, but there is a tendency to connect with other CPUs to take the advantage of the rich corporate database.

GIS technology has special characteristics that favour implementation in a distributed model, rather than centralized model or stand-alone model. These include the following:

- There is a need to support both graphic and tabular processing. Mainframes are less well prepared to handle the greater demand for terminal I/O (input/output) required by graphics applications. Since a GIS does both graphics and tabular processing at the same time, it is impossible to tune a single machine to optimally handle both applications.
- There is a need to support user-friendly, window-oriented interface software (GUI—graphical user interface). The specialized terminals and software required to support this style of user interaction require dedicated processing power. A mainframe that must support

several other applications and possibly hundreds of users usually does not have the resources available.

- The *open-system approach* means the integration of industry-leading hardware and software products into a customized solution that is ideally suited to GIS applications.
- GIS technology is in the mainstream of workstation development. GIS software makes extensive use of workstation processing power to drive graphic display and window-oriented user interfaces. Distributed GIS technology makes full use of the network capability built into workstations. Also, GIS technology benefits from the development that appeared first in the workstation environment.

X-Terminals in a GIS environment

X-terminal technology is being used in organizations for low-cost queries and 'occasional use' applications, particularly in GIS configurations with a large number of users. In order to take advantage of the vast, untapped computing power available in modern workstations, several X-Terminals can, via the network, log onto a workstation and concurrently run their own programs (LIC 1993). This provides a very cost-effective means of making graphic screens available to many users for GIS attribute data capture, data validation and viewing, without the need to purchase additional workstations. It could be cost-effectively configured for spatial data capture as well (Fig. 2.20).

2.4 GIS standards

GIS industry, in some sense, is a most unregulated industry. There are very few standards in the world that apply to many aspects of this industry. GIS standards can be defined as a set of procedures, requirements or formats that must be adhered to in order to satisfy a set of specified criteria. A GIS depends on standards in the wider computing community, as well as

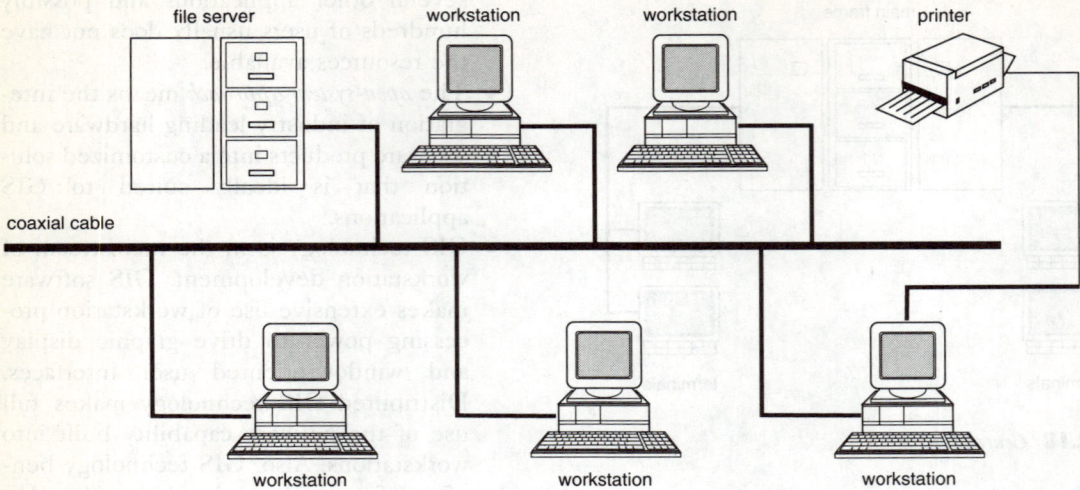

Fig. 2.19 *Distributed model*

specifying standards of its own. The standards of a GIS are coming from two major sources: one is the governing body in a country, such as the National Standards Agency, and the other is from industry (which adopt some de facto standards).

Coleman and McLoughlin (1994) list a number of potential advantages that are offered by standards:

- wider access to GIS at all levels within an organization
- easier exchange of data within and between organizations
- easier integration of a GIS with other systems in an organization
- transparent data exchange between equipment on a network where multiple systems from different vendors may be in use
- reduced employee training times in initial and subsequent applications
- emphasis by vendors on new applications rather than on supporting proprietary manufacture of different hardware vendors
- more stable production environment and reduced dependence on individual vendors

The most important standards associated with geographic information systems are the data exchange standards, software standards and hardware standards.

2.4.1 Data exchange standards

Some vendors have developed their own exchange formats that suit their particular products and that, due to market share or some other factor, have been adopted as 'industry standards'. Some others are supported as official standards.

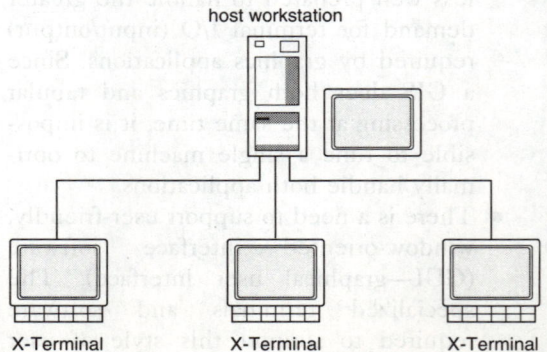

Fig. 2.20 *Configuration of X-Terminals*

American Standard Code for Information Interchange

ASCII was first created in 1966 by ANSI (American National Standards Institute) and subsequently adopted as the basis for an ISO (International Standard Organization) standard. The equivalent Australian Standard is AS1776.

ASCII is intended to define the digital representation of alpha-numeric data, providing for 256 characters.

Street and Rural Address Standard

This standard is intended to facilitate the exchange of digital address data by providing a data dictionary that defines the components of a street address.

Land use classification

A draft Australian Standard titled 'Geographical Information Systems—Land use classification' has been circulated and a similar standard exists in New Zealand.

The Australian draft proposes a six digit code: four digits for the basic code and an additional two for an auxiliary code. The basic code uses the first two digits for a broad category, the next digit for subclassifications and the final digit for further refinement.

2.4.2 Software standards
User interface

The user interface is the means by which users interact with the GIS and provide the look and feel of the GIS. Adoption of standards in this regard can, therefore, provide substantial savings of cost and effort in training, as staff move from organization to organization and system to system.

The majority of geographic information systems provide a GUI. A GUI predominantly takes on one of the two of industry's standards: *OSF* (Open Systems Foundation)/*Motif* and *Openlook*.

X-Windows is a hardware-independent operating system, with a graphics standard designed to operate over a network, or within a stand-alone machine. X-Windows allows total portability of a graphics application between various platforms, provided an X server is available for the required platform.

Database query

For GIS a flexible database query language, which supports ad hoc queries, is essential.

For systems based on the relational database model, standardization is provided through Structured Query Language (SQL). SQL is a non-procedural language that allows the operator to perform tasks by specifying the desired end result, rather than the operations required to achieve it. From a GIS perspective, a major limitation of SQL as a standard is its lack of graphical and spatial query commands.

2.4.3 Hardware standards
Operation system

A portable operating system allows an organization to run on multiple hardware platforms, or to upgrade one platform to another, instead of making a commitment to a particular vendor for an extended period of time. The bulk of GIS products on offer are designed for RISC (reduced instruction set computer) workstations or Intel (or equivalent) based PCs.

In the *workstation* market, UNIX is becoming the predominant operating system. UNIX vendors have recognized the benefits of standardization, and it is likely that soon there will be less than a handful of significant variations of UNIX on offer.

In the past, the PC platform has been completely dominated by Microsoft's DOS. Now OS/2, Windows, Windows NT and adopted UNIX can be found in many PCs.

It can be anticipated that parallel processing technology will mature and will spawn a new generation of operating systems.

Networking standards

Computer networks permit computers to communicate with one another to share access to peripheral devices. In order to transmit data over a network, it is usual to divide a file into smaller units called *Packets*. Packets consist of part of the original file contents together with appended protocol headers.

Network protocols:

- handle the exchange of packets
- ensure packets reach the correct destination
- detect and correct errors in packets
- reassemble packets into the same form as they were originally prior to the transmission.

Local area network

A LAN is generally limited to a few kilometers of cable. The main LAN protocols are standardized by the IEEE (Institute of Electrical and Electronic Engineers) or ANSI.

Ethernet (IEEE 802.3)

Ethernet was originally developed by Xerox, and a slightly different version was standardized as IEEE 802.3. Ethernet typically uses coaxial cable (thick or thin) to connect devices, but may also use fiber optic or twisted copper pair. It is the most common form of LAN and can achieve transfer rates up to 10 megabits/second. Future versions of the standard will support 100 megabits/second transfer rates.

Packets transmitted onto the bus (cable) visit all nodes (stations) on the network until they are accepted by a node that has an address that matches the *packet header address*.

Any node can transmit if the bus is idle, but if more than one node transmits at the same time, their signals 'collide', which can cause the data to be corrupted. When a collision occurs, each station waits for a random period of time before transmitting.

Token ring (IEEE 802.5)

IBM was the main instigator/supporter of the *token ring concept*. It is fast catching up with Ethernet as the most popular form of network. Twisted pair cabling is used to connect nodes into a physical ring, typically operating at 4 megabits/second, but as high as 100 megabits/second (Fig. 2.21).

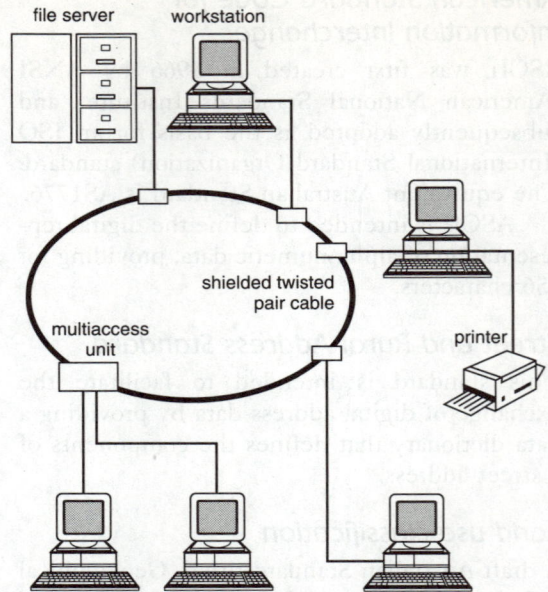

Fig. 2.21 *Network—token ring*

Like Ethernet, all packets visit all nodes until they are 'claimed'; however, nodes may only transmit if they hold a special packet called the *token*. After transmission of the data packet, the token is transmitted (released) and is seized by the next node on the ring waiting to transmit, thus ensuring that no collisions occur.

Fiber Distributed Data Interface

Fiber Distributed Data Interface (FDDI) is an ANSI standard that is conceptually the same as token ring, with changes to suit FDDI's transmission rate of 100 megabits/second over fiber optic cable. FDDI is most commonly used as a backbone onto which other lower speed LANs can be connected, and can extend up to a ring length of 100 kilometres (62 miles) (Fig. 2.22). The high speed of FDDI is particularly suited to the graphic data common in GIS applications.

Wide area network

A WAN allows a connection mechanism across a city or a country, or even between countries. Because of their extent, they are typically based on the public telephone network. Australian

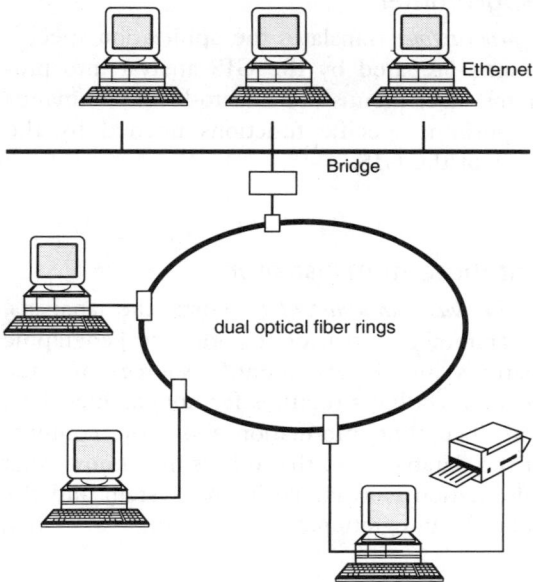

Fig. 2.22 *Network—FDDI*

The purpose of each layer is as follows (LIC 1993):

Layer 7:	Provides a user interface to the lower levels
Layer 6:	Provides data formatting and code conversion;
Layer 5:	Handles coordination between processes
Layer 4:	Provides reliable end-to-end transport
Layer 3:	Sets up and maintains connections;
Layer 2:	Provides reliable data transfer between two network devices
Layer 1:	Passes a 'bit' stream between terminal and network

Telstra offers ISDN (Integrated Services Digital Network), which is capable of 64 kilobits/second, or FASTPAK, which is currently offered at up to 10 megabits/second but which is ultimately capable of far greater speeds. Both ISDN and FASTPAK can be used to link the LANs as previously discussed.

2.4.4 General standardization issues in networking

Interoperability

The interoperability refers to the ability of users on different computers, linked by different networks, to share the same files and peripheral devices. The standard that has the greatest impact on interoperability is the *Open Systems Interconnection* (OSI) Reference Model, established by International Standards Organization (ISO). The OSI Model defines seven protocol layers, which can be considered quite separately, but which have well-defined interfaces between the layers. This allows developers to ensure that their product can communicate with other products in adjoining layers (Fig. 2.23).

2.5 Human resources

A GIS is a front-running and interdisciplinary technology, and it requires qualified human resources in several areas of operational levels. The general human resources categories are listed below. However, depending on the size of the computerized information system, several different computer installations can perform interactively.

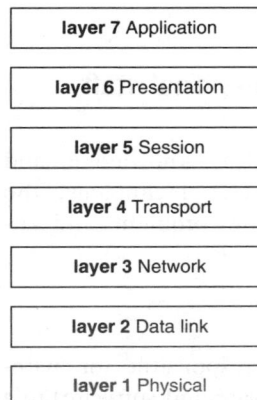

Fig. 2.23 *Open Systems Interconnection model of ISO*

2.5.1 Operational staff

End user

Those who make the system produce the benefits that were originally anticipated during the planning and evaluation stages of the GIS project (that is, using the capabilities of GIS technology to meet the geographic data-processing needs of the functions) are called *end users*.

Cartographer

The design of map displays should be clear and understandable, and deliver the intended message. The *cartographer* also assists in the development of standard map symbols and establishes standard map series for general distribution.

Data capturer

The *data capturer* converts map information into digital form. This activity requires people who can work for long periods of time with the digitizer/scanner or the workstation screen.

Potential users

There are *potential users* who will need to use the GIS technology, because a change in the organization's function and the suitability of GIS, that is, an expansion of the database, a higher degree of database integration, and new functions added in a GIS.

2.5.2 Technical professional staff

Analyst

Utilizing specific technical knowledge and experience in applying GIS technology, the analyst solves particular user problems and satisfies their information needs.

System administrator

A *system administrator* is responsible for maintaining the system (hardware and software) in a continuous operational mode, responding to and solving problems as they occur.

Programmer

A *programmer* translates the application specifications prepared by the GIS analyst into programs, user menus and macro-level commands to perform specific functions needed by the users of the GIS.

Database administrator

A *database administrator* assists the analysts, programmers and users to organize geographic features into layers, identify sources of data, develop coding structures for nongraphics data, and document information about the contents of the databases, so that others may know what information is available in the system and the database integration.

Super-operator

A super-operator knows all the capabilities of the GIS hardware and software, and can utilize them to produce specific products needed by the users and a joint coordinating training program.

2.5.3 Management personnel

Manager

The *manager* is responsible for the adequate daily performance of the LIS/GIS project implementation team, and manages the output/production as required by the organization.

QA coordinator

The QA coordinator provides quality assurance to the team with other roles, such as end user in the conversion process design, the data acceptance plans, and the quality control procedures. This person is usually responsible for managing the output of the final product to ensure it meets the conversion specification and data acceptance plan.

2.6 A simple guide to the evaluation of GIS software

GIS technology is becoming more user-friendly, more extensive and analytical in its applications, more miniaturized and more cost-effective. Massive and costly mainframes of 10 years ago are giving way to engineering workstations and personal computers, performing more interactive mapping and analytical functions at greatly reduced cost. Although it is necessary to have a fully functionalized GIS software, it is not necessary to chose all the software from the same vendor. Versatility and functionality of a *geographic information system* can be judged by the following criteria:

1. **Friendly user interface and development environment**

   Specialists today recommend that the 'system' contain features of a friendly *window and multilingual* environment or Native Language Support (NLS), with on-line help, user-generated macros, user customizable menus and 3GL/4GL language for applications development.

2. **Multichannel data transformation and exchange**

   Specialists also recommend that a 'system' may need to support major data structures, such as topological vectors, nontopological vectors and raster, along with 3-D surface model, Digital Elevation Model (DEM), Triangulated Irregular Network (TIN), and so on.

   Provision needs to be made in some installations that data can be converted between the *coordinates* system, between *projections* and between *vectorial and rasterial* format.

   Spatial data can be imported and exported in the format of industry standards or other major GIS systems. Such standards are American Standard Code for Information Interchange (ASCII), Data Exchange Format (DXF), and Spatial Data Transfer Standard (SDTS). Other major GIS systems are ARC/INFO, Intergraph Modular GIS Environment (MGE), GenaMap, AutoCAD based GIS, SPANS and MapInfo (see Chapter 7 for a description of the several software capabilities).

3. **Relational data management capability**

   The 'system' often needs to have the function of an internally Structured Query Language (SQL) or have a database integrator to interact dynamically with some major relational database management system (RDBMS), such as DB2, Oracle, Ingres, Infomix, Sybase, Rbase, Dbase and Foxbase.

4. **Universal hardware, peripheral configuration**

   In some cases, the GIS to be installed would need to be fully integrated with an existing corporate system. The 'system' should be designed that has compatibility with, and can be run through, the network on different platforms, ranging from PCs, workstations to miniframes/mainframes, which use DOS, Macintosh, various UNIX, and even VMS (Virtual Memory System) operating systems to meet various users' needs. The system should provide that data can be captured by scanner, digitizer, photogrammetric station, data recorder, global positioning system (GPS), and so on, and can be sent to major brands of printer, plotter, film recorder and so on.

5. **Comprehensive analytical functionality**

   (a) *Map/Polygon analysis function* provides for an overlay of many different layers or objects of spatial information at the same time. If necessary, reclassify the map. The map/polygon function has the ability to conduct contiguous/ nominal overlays, non-contiguous/nominal overlays, contiguous/ attribute overlays and noncontiguous/attribute overlays, and can dissolve the boundary after overlays have been transacted.

(b) *Surface analysis* calculates surface slope, compass aspect, area of surface, and distance along a straight line and arc. Also, it has features of generation of cross-sections and long sections, volumetric cut and fill, specific interval contouring, wireframing, and so on.

(c) *Network, adjacency and proximity analysis* features routing and minimum path along network, spatial adjacency search, nearest neighbor search, address matching, proximity/ weighted proximity analysis, and buffering near a point or along an arc or around a polygon.

(d) *Digital image analysis* has capabilities of:

  (i) image preprocessing—radiometric corrections, sensor corrections merge data sets and geometric corrections

  (ii) image enhancement—filtering, user definable filters, contrast stretch, color domain conversions, density slicing, histogram, histogram equalization and mosaicing

  (iii) image extraction—principal components analysis, band ratios, supervised classification and unsupervised classification

  (iv) expert system applications

6. Data representation

This function has capabilities of:

(a) features drawing and editing

(b) annotation

(c) providing tools for adding cartographic elements

(d) supplying from a library of patterns of symbol, line, area and text

(e) producing tables, statistic graphics together with a map

7. Application modules

It is also very important that the vendor or the third party supplies the particular application modules to save the time for application development, such as natural resource planning; network design and analysis capability for pipelines, electricity and road network; land titling and recording; and automated procedures relating to spatial and aspatial data manipulation, as well as dynamic monitoring and control applications.

CHAPTER 3

Applications of a GIS in planning and decision making

3.1 Principle of using a GIS for planning and decision making

Planning is a normal and pervasive part of human existence. Planning has three characteristics: first, it must involve the future; second, it must involve action; and third, there is an element of personal or organizational identification or causation. The future course of action will be taken by the planner or someone designated by or for him/her within the organization. Futurity, action, and personal or organizational causation are necessary elements in every plan (Lebreton & Henning 1961).

The classical planning process is enumerated as follows:

- problem identification (awareness of need)
- goal setting (statement of objectives and establishment of a work program to prepare appropriate plans)
- data collection and analysis
- refinement of goals
- development of alternative plans and/or policies (designed to achieve goals)
- evaluation of alternatives (determination of probable effects, both good and bad, and the ease or the difficulty of implementation)
- adoption of preferred plans and/or policies
- implementation of plans and/or policies

- monitoring and evaluation of results (alerts to progress toward goals and/or danger signs calling for course correction)
- feedback (recycling the planning process as necessary to meet emerging circumstances)

Planning and decision making in government could be classified into various levels: local, regional, provincial/state, inter-provincial/state and national. Most of the issues are addressed within the arena of urban and regional planning.

Planning and decision making at local level are traditionally concerned with land use zoning, regulation of the land subdivision process, urban reconstruction, public housing, local environmental protection, and the provision of local parks and public works. This involves the preparation of local comprehensive plans and capital improvements programs, in which public action, supporting orderly physical development of the local community, is mapped out and budgeted, and explored in detail as the accepted means toward effective exercise of recognized local government powers.

Regional planning is not just local master planning on a grander scale. It has its own characteristics, features, topical interests and processes. Regional planning, in particular, is different in that it is related directly to the exercise of governmental authority on a region-wide basis. Most regional planning organizations, whether of the large multistate/provincial variety encompassing the multiple local governments in a single metropolitan area, or any other contiguous community of social contacts in the daily working and living environment, are inter-governmental mechanisms.

The broad areas of interest in planning include *economic development*, *environmental protection*, *resource planning*, *infrastructure provision* and *social welfare improvement*. Government at different levels plays different roles in each field.

A GIS is an effective tool for planning and decision making. A GIS is best defined as a decision support system involving the integration of spatially referenced data in a problem-solving environment (*GIS World* 1993).

A GIS is an effective tool for urban and regional planning, particularly in the era of more concern about the environmental impact of economic development. The Brundtland Report, *Our Common Future*, prepared by the World Commission on Environment and Development for the General Assembly of the United Nations in 1987, outlines a number of development approaches, many of which would be greatly assisted by the application of a GIS. The report calls for a sustainable development strategy, which ensures human progress for today without bankrupting the resources of future generations. The emphasis is on the need to merge environment and economics in decision making. Due emphasis should be given to the impact of development projects on the environment to see whether or not they are ecologically and economically sustainable. Geographic information systems have the capacity to provide powerful assistance in supporting the objectives of the Brundtland Report and in coping with many of the issues identified. Geographic information systems can enable the monitoring and analysis of both local and regional major environmental issues raised in the report, such as deforestation, industrial pollution, soil loss, land degradation and other sources of deterioration of the national resources base. They can also assist with monitoring urban expansion to avoid harming the environment.

As most geographic information systems in the developing countries are regional and *resource and environment based*, they are especially useful for implementing the sustainable development strategy (Yeh 1991).

Both in-house geographic information systems developed for planning and decision support, and the application of commercially available geographic information systems have arisen in the past few years. Compared with in-house developed geographic information systems, commercial GIS packages normally are less expensive and more flexible, compatible and accessible.

GIS application to planning, generally, could be classified as shown in Figure 3.1.:

- *Inventory*—Store a complete description of the study area and account for changes in the physical world.
- *Analysis*—Explain and exploit the existing database components to produce high-quality information.
- *Modeling*—Using most of the existing data, control some of the variables, and forecast the overall change.
- *Presentation*—Generate static or dynamic maps, tabular summary and statistical graphics.

Regardless of their specific technical details, GIS applications in planning may be located along a continuum in terms of their capabilities and the kinds of problems they address.

Most LIS/GIS applications have been directed at tasks towards the left-hand side of Figure 3.2. This is also the area of the application of computer mapping. Emphasis is on answering questions of the 'what is' and 'where are' varieties which may simply involve the production of a thematic map of a single item. More commonly, however, questions encompass a combination of items, for example: Where are the places with a particular set of attributes? What are the characteristics of a particular place? How much of a particular attribute, or combination of attributes, is present in a particular location? Typically, there is only one answer to these questions—essentially a description in map and record form of the existing situation. The capacity for GIS to address these kinds of questions in *planning* is now well documented, the answers to which are important for what may be thought of as the discussion-promoting role in decision making and policy formulation. At this level of application, GIS is used primarily for description and to provide users with *inventory* information.

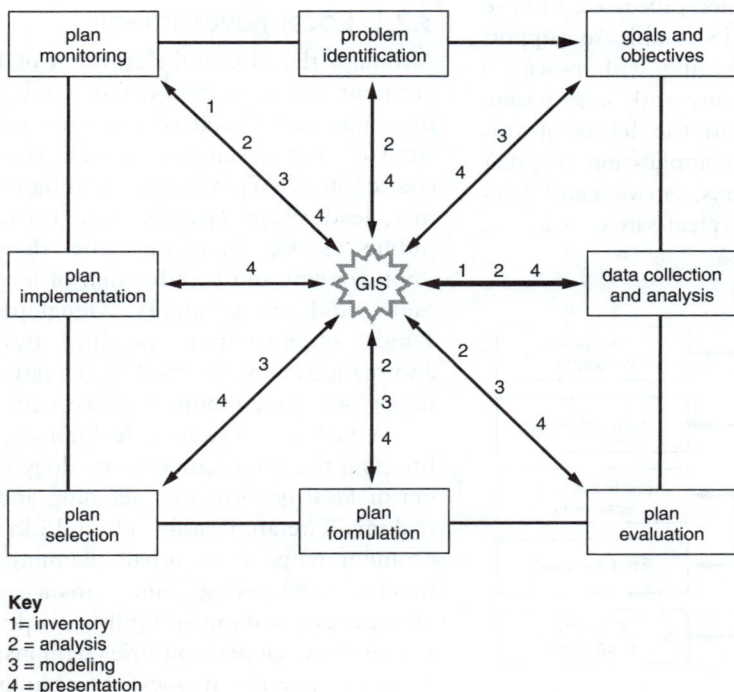

Key
1 = inventory
2 = analysis
3 = modeling
4 = presentation

Fig. 3.1 *A GIS in planning practices*

Planners also need *process information*, however, which is information describing change over time in attributes, their distribution and their interrelationships. This may be obtained from repeated inventories as part of the monitoring process, or by analytical modeling, to generate predictions and scenarios of future states in a given spatial system— this is a much more difficult task, but one which is an integral part of the policy-making process. Given the present state of GIS technology, their use so far has been much more limited in addressing these more complex problems for which the central concern is one of invention rather than inventory. By the term 'invention' is meant the ability to answer questions of the 'what if' and 'where' variety: these are questions which are fundamental to spatial forecasting, evaluating alternative policies and the formulation of new plans.

GARNER & ZHOU 1990

There is a huge range of GIS applications in various planning and decision processes. But the principle is the same: GIS is able to support planning and decision making with powerful inventory, analysis, modeling and presentation functions. It is far beyond the limits of this guideline to discuss each application in great detail, but some applications, software and data-base requirements in typical areas will be covered in general.

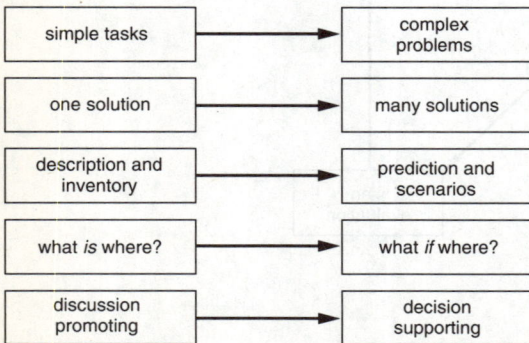

Fig. 3.2 *GIS applications in planning and decision making*

Since planning and the decision-making process in local government is comprehensive, and physical, GIS applications range from land use planning to facility management.

The use of a GIS at other levels of government is in departments, sections and project-oriented activities. Different departments may realize the importance of data sharing, but planning and policy making issues, GIS application, project implementation and other issues are invariably limited in certain areas. Such issues are mineral resource management, environment protection and monitoring, social-economic planning, and facility management.

3.2 GIS applications in planning and decision making

3.2.1 Local government

Although the size and definition of local government may vary from country to country, the functions that a local government performs are similar. For example, Local government's charter in NSW, Australia, highlights community leadership; custody and trusteeship of public assets; facilitating the development, improvement and coordination of local government; and accountability. Geographic-related Lands administration, facilities development and management are the key operation areas as far as local government is concerned.

A GIS is a dynamic technology, which is bringing the information technology into a new era of local government planning and decision making. The applications of a GIS in local government range from urban planning, environmental monitoring and management to emergency planning and public works. The situation in developed countries demonstrates that a GIS is popular in local government (Table 3.1), and local government is one of the biggest GIS user groups overall.

Table 3.1 *Geographic information systems in local government from selected countries*

Country/ Region	Operation stage (%)	Planning stage (%)
Australia		
NSW	46.3	30
VIC	35.7	na
WA	26.1	na
New Zealand	54.7	na
USA	68	17
UK	16.5	52.8

Source: AISIST 1993b

Typical applications

Typical applications are:

- land-use inventory
- zoning
- transportation plan
- general planning
- facility and utilities planning
- growth monitoring
- land parcel mapping—cadastral maps— (DCDB—Digital Cadastral Databases)
- census mapping
- community development
- utility mapping
- engineering

GIS data–Brisbane City Council

An example in Australia of GIS data from Brisbane City Council, 1994, is:

- bikeways
- boundary:
  - federal electoral
  - state electoral
  - wards (local electoral)
- city base map
- bushland
- bus routes
- cadastre
- CBD (central business district) street lights
- community facilities
- contours: 1 metre (1 yard), 5 metres (5 yards), 10 metres (10 yards)

- drainage plan numbers
- environmental status
- flood regulation lines
- flood search flag
- grass cutting
- health
- heritage listings
- kerb parking
- parks
- park facilities
- park lease boundaries
- pavement marking (maintenance)
- prescribed purpose site
- road planning
- sewerage reticulation
- storm water
- storm water maintenance
- streets
- street furniture
- survey marks
- survey traverses
- traffic counts
- traffic signals
- traffic signs
- vegetation
- waste collection run
- waste contract boundaries
- water reticulation
- waterway plan units
- wetlands
- works district boundaries
- works requisition search
- zoning

Software

GIS software used in local government is mostly unpredictable. In-house developed software has a significant share in some countries. Commercial GIS software ranges from desktop mapping software to fully customized *large scale, multipurpose digital cadastre database*. The size of population and number of land parcels in local government are the most important factors for local governments selecting their software.

Requirements

Generally, a local government requires a multi-functional GIS package, including data integration; network analysis; surface analysis; spatial data query; and excellent function in accessing other corporate data, and so on.

Typical software

Typical software is:

- USA:
  - Arc/Info
  - PC Arc/Info
  - Intergraph MGE
  - MapInfo
  - GDS
  - GeoVision
  - MapGrafix
  - Genasys
- UK:
  - Arc/Info
  - Alper Records
- Australia:
  - Mapinfo
  - GenaMap
  - Arc/Info
  - Datamap

3.2.2 Natural resource management

GIS applications range from simple inventory and query of spatially located objects, to map analysis based on geographic operations, to the support of complex spatial decision-making systems. A GIS can put this capability at the fingertips of decision makers at every level of an organization from the field person to the policy maker.

Typical applications

Typical applications (Mahoney 1993) are:

- forest inventory
- forest modeling
- production estimation and management
- access planning
- fire risk analysis and planning
- monitoring the change of ecosystem

- monitoring acid deposition and other environmental hazards
- impact analysis on how economic, meteorological, hydrological and other processes interact with geographically disposed natural resources

GIS data

GIS data is:

- elevation
- forest cover
- soil quality
- land classification
- water and river resources
- coastal and marine resources (as applicable)
- trends of soil erosion
- crop production
- animal grazing capacity
- geological strata and mineral potential
- effect of urban built-up areas on agricultural land
- natural hazards, such as bush fires, cyclones, malaria or other fevers

Software

Requirements

Requirements are:

- significant polygon overlay function
- integrate spatial data in raster and vector
- image processing
- surface modeling
- 3-D visualization

Typical software

Typical software is:

- GenaMap
- Arc/Info
- ERDAS

3.2.3 Environmental planning

Environmental planning is defined as a planning process through which environmental considerations are incorporated into socioeconomic development. It is only through this type of planning that environmental matters are comprehensively and completely integrated into the development process, thereby securing sustainable development. Measures such as environmental institutions and legislation, education and training, public awareness and participation, and the application of environmental technologies are the elements necessary to facilitate environmental planning and management.

Tools for environmental planning

The tools for environmental planning (ESCAP 1990) are the following:

Environmental Impact Assessment

Similar to economic analysis and engineering feasibility studies, an Environment Impact Assessment (EIA) is a management tool for decision makers and planners, who must make important decisions about major development projects. It is also a process (not a product) through which full environmental considerations are incorporated into project planning, construction and operation.

Environment risk management

Environmental risk management is generally defined as a sequence of four actions, namely:

- hazard identification
- exposure assessment
- hazard assessment (or dose-response assessment)
- risk characterization

Environmental accounting

Two principal elements of environmental accounting need to be discussed. The first element relates to the measurement of environmental quality and resource stocks. This is an essential intermediate step in the preparation of environmental responsive national accounts. The second element requires these physical accounts to be related to monetary aggregates of resources stock and flow.

Environmental monitoring

Environmental monitoring can be defined as a process for repeated observations and measurements of physical, chemical and/or biological parameters of single or plural elements of the environment at specific places and intervals with particular objectives. It is becoming clear that these need to be supplemented with the monitoring of socioeconomic parameters.

Environmental database

A systematic information system, which might be referred to as an 'environmental database', is a valuable aid to decision makers and planners in formulating and evaluating comprehensive and effective environmental policy. This system covers the natural as well as the human environment, including a wide range of human activities, natural events and environmental impacts.

Environmental standards

All environmental planning must be related to objectives that may be termed 'environmental standards'. These standards are minimum quality-of-life values acceptable for human well-being.

Typical applications

Typical applications are:

- providing effective analysis and modeling tools for an EIA
- air/water quality modeling and monitoring
- making an inventory of environmental data
- risk evaluation
- environmental sensitive zone mapping
- environmental impact study
- coastal management

GIS data

GIS data is:

- administration boundary
- topographic data
- vegetation data

- drainage/shorelines
- land condition
- hydro-geology
- soil type and texture data
- road center line
- precipitation and evaporation
- temperature

GIS software
Requirements
Requirements are:

- significant polygon overlay function
- proximity analysis
- buffer generation
- 3-D visualization
- seamless link with other modeling packages

Typical software
Typical software is:

- Arc/Info
- GenaMap
- Intergraph MGE

3.2.4 Emergency planning and management

Natural hazards and hazards caused by human intervention kill tens of thousands of human beings, and damage billions of dollars worth of property each year around the world. The following is a short checklist of the hazards encountered:

- Natural hazards are:
  - riverine floods
  - forest and bush fires
  - cyclones
  - landslides
- Hazards caused by human intervention are:
  - road accidents
  - building fires and collapsed buildings
  - chemical leaks
  - oil spills

The process of emergency management could be well represented as a *disaster cycle* (Johnson & Granger 1994). There are several equally important phases in a disaster cycle:

- hazards analysis
- vulnerability analysis
- mitigation
- preparation for alleviation of hazard
- prediction and warning
- response
- recovery from disaster

Since each phase is geographically related and requires geographic information and spatial analysis and presentation, a GIS is very valuable as a tool for planning and decision making in emergency management. It not only can assist in damage assessment, but also can provide critical information during the response phase.

Following are a few examples:

- providing an incidence map and thematic map during the hazards analysis phase
- providing a map and tabular report for shelter inventory analysis; identifying location, address and characteristics of shelters within a county; and providing a map for multihazard analysis showing an overlay of, for example, a potential flood zone or a hazardous material site, with a calculation of the distance to the nearest residence during hazard vulnerability and mitigation phase
- evacuation routing during the prediction and warning phase
- damage incident reporting and emergency vehicle routing during the response phase
- damage assessment and project monitoring during the recovery phase

Typical applications
Government departments

In government departments (primary resources, planning, transport, and so on) applications include a general understanding of the hazards problems, hazards analysis, monitoring, prevention and mitigation, enhancing emergency response capabilities, relocation of the relief resources, damage assessment, and so on.

Ambulance services

Ambulance services applications include route planning and 'optimization'.

Fire brigade

Applications for the fire brigade include fire case analysis, routing optimization, and detailed property and utility information.

Police

Applications for the police include criminal trend analysis, road accident analysis, accident site location, re-routing traffic, VIP route planning and patrols route planning.

Insurance and finance

Insurance and finance applications include hazard analysis and mapping, insurance grade mapping, policyholder mapping, risk analysis and damage assessment.

GIS data

GIS data is:

- street network
- topographic data
- demographic and economic data
- rasterized aerial photography

Software

Requirements

Requirements are:

- access to other existing digital textual databases
- integration of global positioning system (GPS) and communications technology with GIS
- effective data query function
- comprehensive spatial analysis function, such as network analysis
- decision support procedures for a particular emergency situation
- extensive use of aerial photography and remote sensing

Typical software

Typical software is:

- MapInfo
- Intergraph MGE
- Arc/Info
- ERDAS
- ATLAS

3.2.5 Mineral resource management

Geographic information systems and remote sensing (satellite and airborne) are playing an increasingly important role in multilevel multistage planning and decision making in mineral resource management. At national/state/provincial level, it could be used for inventory, modeling and mapping mineral resources. While on *site*, it could be used as:

- mineral and oil exploration
- assessing preproduction conditions
- monitoring external impact of production
- measuring the production rate
- rehabilitation of production sites
- site utility management

Typical applications

Mineral resource management

Typical applications are:

- inventory
- reservoir estimation
- exploration and mining planning
- reservoir mapping

Exploration phase

In mineral and oil exploration, *information subsets* include rock type; rock age; oxidation state; metamorphic grade; presence or absence of faults and other structures; distance from granite; whole rock and trace element geochemistry of the rocks; magnetic, gravity and radiometric signatures; and so on. A GIS not only allows for different layers of information to be overlaid in a computer graphics environment, but also enables elements of the different layers to be blended or interrogated, one against the other.

Preproduction phase

A GIS is an important tool in Environmental Impact Assessment, especially for feasibility studies of mineral and oil exploitation projects. With GIS, solutions to such problems as balancing mining and forestry, balancing mining and agriculture, industrial facilities siting, site selection for waste dumps and hazardous materials removals can be facilitated.

Production phase

GIS can assist in resource valuation, tax reporting and land management.

Utility management

Mines and/or oil fields have complicated facilities. The role of AM/FM/GIS can be important, just as it is becoming so in local government or any *utilities* industries.

Rehabilitation phase

After mining or oil production, most governments require the companies to establish long-term forest ecosystems, with measurable leaf areas, hydrological balance and sustainable timber production. GIS-related monitoring and change-detection techniques can be used for these purposes.

GIS data

GIS data is:

- cadastral data
- topological data
- road and access
- water reserve and water system
- vegetation
- reserve information
- historical site information
- geological data

Software

Requirements

Requirements are:

- capacity to handle massive geophysical, seismic, satellite and airborne data

- statistical graphic generation functionality
- extensive use of remote sensing and GPS technology
- 3D DTM, and 2-D and 3-D seismic modeling

Typical software

Typical software is:

- Arc/Info
- Intergraph MGE
- IDRISI
- Geo/SQL
- ERDAS
- ER Mapper
- GenaMap

3.2.6 Transport

GIS technology is now extensively used in every corner of transport planning, from transport marketing to the large-vehicle fleet navigation. The advent of inexpensive, easily accessible GIS technology has placed the transport planning, analysis and modeling processes in a new light.

Typical applications

Fleet monitoring and navigation

The management of a fleet of vehicles combines mapping, communications and position location of vehicles, such as locating the fleet position in a vehicle environment, monitoring the fleet speed and rerouting the fleet in the case of road accidents and other reasons.

Network analysis planning

A *transport planning and network analysis* can cover the total transport system, such as city bus routes and timetables, city train routes and timetables, and volume of persons travelling at any one time during the day and night.

Routing optimization

Routing optimization could involve describing customer locations for prospecting, shipping and billing, and identifying optimum routing options for different commodities with varying constraints.

'Transport demand' modeling

Deriving data for small homogeneous areas from industry, population and employment statistics, a *transport demand* model can be set up to determine the requirements of the entire network needed to meet current and anticipated transport demand market opportunities. A GIS can assist with locating stations, arranging routes and timetables.

Accident analysis

Accident analysis requires the correlation of a number of explanatory roadway and environmental variables, such as sidewalk condition, roadway geometry, weather conditions, traffic volumes, signals and lighting. An established way to associate these variables is through a common geographical referencing system.

Data set

Data set is:

- road network
- street address
- topographic data
- real-time locational data
- station data
- transport inventory data

Software

Requirements

Requirements are:

- GPS, and communication units at mobile units
- network analysis capability
- seamless linkage to other analysis models

Typical software

Typical software is:

- MapInfo
- Arc/Info
- GDS
- GenaMap

3.2.7 Public utilities

A GIS is quickly becoming an integral information system component in all utility applications: electric, gas, water, roads, telecommunication, storm water, sewers, TV/FM signals and others. Today, most utilities consider a GIS a strategic corporate asset rather than an isolated, proprietary system, useful only to one or two departments.

A GIS represents a major financial commitment and a fundamental change in the way utilities do business. These organisations are looking toward the significant benefits that accrue from GIS technology, both in *qualitative* improvements in planning, decision support, response to customers, regulatory requests, corporate images, standardization of methods, graphics displays, maps and symbols, and value of the database over its life span, and in *quantifiable* economic results.

Typical applications

Typical applications are:

- asset inventory
- network analysis

A GIS allows planners to alter and enhance the facilities model to maximize system efficiency, or to project the results of potential upgrades to the facilities system. Network and fault analysis includes load flow analysis, short circuit calculation, leak analysis, economic calculation, and so on.

GIS data

Base map data

Base map data (within certain size of buffer of facility) is:

- cadastral database
- topographic data
- local government administration boundary
- roads
- building
- others

Asset data

Asset data (electricity for example) is:

- substations
- duct network
- access holes
- cable
- splices
- fibers
- termination points and others

Software

Requirements

Requirements are:

- excellent data conversion function and/or software
- significant data-storage capability
- superb map joint function
- software with both GIS and engineering design/drafting functions
- network analysis capability
- strong data adjustment function
- network and fault analysis module

Typical software

Typical software is:

- GDS (ARC-NET in Australia)
- Intergraph MGE
- Arc/Info
- Geovision
- GenaMap
- Small World

3.2.8 Socioeconomic development

Population is constantly changing in its size, rate of growth and characteristics. The way in which people earn their livelihood, how they group themselves into families and households, where they live and how often they move house also change over time. These demographic shifts comprise one of the elements shaping the demand for particular types of infrastructure, utilities, goods and services. They also shape the population's impact on the environment.

A GIS is able to act as an integrator of vastly different datasets that are spatially referenced. The application is countless, from thematic mapping of major socioeconomic index to the forecasting of socioeconomic change.

GIS data–example in Australia

Spatial data

Spatial data is:

- statistical local area (SLA)
- legal local government area (legal LGA)
- statistical division
- federal electoral division
- area in hectares
- latitude and longitude of its centroid
- an indication of the comparability of the area with the two previous censuses
- a selected range of preliminary and final counts

Census data

Census data is:

- age
- birthplace
- citizenship
- dwellings
- education
- ethnicity
- family
- household
- housing costs
- income
- industry
- internal migration
- journey to work
- labor force
- language
- marital status
- motor vehicles
- nature of occupancy
- nonprivate dwellings
- occupation
- offspring
- qualifications
- region
- usual residents
- year of arrival in Australia

Other data

Other data (based on the statistical local area) is:

- natural resources
- land use
- topographic features
- road/rail network
- socioeconomic profile
- demography and health
- economics
- business profiles
- labor markets

Software

Requirements

Requirements are:

- excellent data classification and thematic mapping function
- seamless linkage with spreadsheet, and statistical analysis package

Typical software

Typical software is:

- MapInfo
- PC-ArcView
- PC-Arc/Info
- SPANS

3.2.9 Private application: marketing research

The driving force of applications and the development of a GIS for marketing in the 1990s is the potential of the GIS technology. Marketing is the customer, the store, the sales territory, the service center and the distribution center. Companies are continually asking questions that can best be answered geographically and visually: Where are my customers and competitors? Where should I locate my warehouses? Where should I draw the boundaries of sales territories? What are the shortest routes for deliveries?

Typical applications

Distribution

A GIS can help *plan marketing and sales initiatives* and *give its representatives information to help*

retailers to sell. By mapping the location of competitive distributors, it becomes apparent where they have to have their new products represented. Other applications are sales force optimization, advertising and distribution analysis.

Retailing

A GIS can be used for effective mapping of *trade areas* for identifying competitive advantages in site locations, and for merchandising mix retail-based markets, along with shopping mall development.

Finance

Banks are beginning to use geographic information systems to manage assets, explore possible locations for new branches (that is, a real estate program), identify potential clients and increase their understanding of existing bank operations. Geographic information systems have revolutionized the analysis of where business and, consequently, profitability comes from by locating *customer information geographically.* The major application of the system is *catchment area analysis*—an examination of the demographics within the trading area of a service center to determine its business potential. This is now in use by an Australian leading financial institution.

Insurance

Geographic underwriting systems are now in use to provide computerized hazard information classification service, along with risk analysis.

Real estate

Precise and comprehensive information from the realtor has helped to sell property more quickly and conveniently, including purchase, investment and management. Spatial databases can link all the factors, such as demography, facilities, environment, transport, weather, soil quality and hazards history, in the local realtor GIS.

Health care

Applications in health care include sales force workload analysis, and health care coverage analysis.

Direct mail

A GIS can be used to identify potential buyers, linked with demographic survey and analysis.

Advertising

Applications in advertising include audience targeting in radio and television using demographic survey and analysis.

Delivery

Applications include routing optimization and geocoding.

GIS data

Most data is similar to that required for social-economic development. Other data may be required such as street address data and special surveyed data.

Software
Requirements

Requirements are:
- desktop mapping system
- geocoding
- universal polygons and area operations
- effective data and map interaction
- available and affordable spatial and non-spatial data
- low start cost
- seamless linkage to other marketing analysis package

Typical software

Typical software is:
- Arc/Info
- SPANS
- MapInfo
- TACTICAN
- ATLAS

CHAPTER 4

The approach to acquire a GIS

4.1 Introduction

A spatial or geographic information system is rarely a turnkey system. The world is strewn with technological failures. Pearce (1990) summarizes some of the experiences of GIS purchasers over the past decade, as follows:

- The system does not meet the user's expectation.
- The system does not fulfill the business requirements.
- The system cannot be operational within the necessary time scales.
- The system cannot be delivered within the available budget.
- The system does not have adequate performance.
- The system cannot be expanded to cater for future growth.
- The system requires significant development before it is usable.
- The system is not being developed or enhanced by the supplier.

These are hardly faults of the hardware or software, the data or the people; they are faults of the procedures of *project implementation*. There are a few approaches available for GIS implementation in the various business environments that can be considered (Mahoney 1993).

4.1.1 Approaches for GIS implementation
Project-driven approach

The project-driven approach is adopted to *quickly introduce GIS technology* in support of a specific application. It may be used where a *single focused business need* is identified, and a window of opportunity for investment is available. It can be very successful in the short term.

The speed of implementation compromises the evaluation process, and thereby increases the risk that the selected system is not optimal for the purpose. Also, it is not compatible with the longer term objectives, including data integration.

Sometimes, the project-driven approach is used as the initial stage for the full implementation in cases where initial corporate justification has not been thought feasible, few GIS skills are available, and senior management personnel are not convinced of the merits of a more classic approach to evaluation. This methodology can, under some circumstances, lead to a situation where the investment is so small, that the benefits are insufficient to provide the confidence necessary for further investment.

Supplier–partnership approach

The supplier–partnership approach normally arises when the *client enters into a partnership with a supplier*. This approach avoids the need for open tendering, and requires a high degree of integrity from the supplier.

The opportunity for these types of partnership arrangements will diminish as more GIS installations become available on the market, and more local authority applications are developed. The approach places much of the risk with the suppliers, though there is a requirement for a substantial commitment in time and resources from the client involved.

The multistage approach

The multistage approach represents the most rigorous methodology for evaluation and implementation. This approach is most suited to the *corporate* rather than the *departmental* approach. It supports the formulation of a corporate GIS strategy in compliance with the corporate information system strategy. Although there are some variations of this approach, it represents a low-risk approach, but it is a time-consuming and costly option (Fig. 4.1).

4.1.2 GIS implementation

To planners and decision makers, it is very easy to chose the project-driven approach to start the GIS project for the purpose of planning and decision making only. Although it may be very successful politically in the short term, it may end up as a useless system and an unadjustable budget.

> GIS involves heavy investment, and a cautious approach should therefore be taken in the design and setting up of the system. Careful institutional and economic feasibility studies should be undertaken

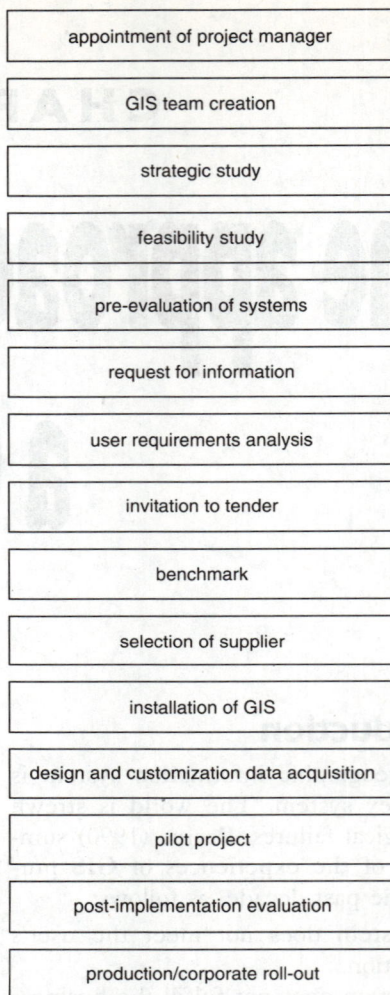

Fig. 4.1 *Multistage approach*

before a start is made with a GIS. It is difficult to justify a GIS for regional planning alone, and urban and regional planning will remain as one of the users rather than the owners of the system. The need in the developing countries is to improve the institutional arrangement and application aspect of GIS rather than the technology. Planners need to develop skills and methods in using GIS which will be increasingly available in the developing countries.

YEH 1991

Ineffective systems are those designed only for operational activities, or only for management or policy purposes ... The latter require data to be created outside the normal operating activities, and do not have the support they need when workload demands from other tasks increase, or when budgets get trimmed.

Some cities or departments within cities have implemented a geographic information system to solve a particular problem, or address a specific issue related to public policy, and have collected pertinent information, digitized maps from various sources, and analyzed the data for the project or program. Later, when the problem or issue no longer is important, the system could not be used for other purposes, because the data were not kept current and the analysis programs were not generalised for other uses.

HUXHOLD 1991

GIS project implementation is not just a transaction process from manual geographic information processing to a digital automatic GIS, or from a single functional CAD/CAM system to a multifunctional GIS. It will also permit a change from a segregated system to an integrated corporate-wide system. The focus of managing GIS technology acquisition requires a change from a single project, single department and single purpose of planning and decision-making implementation to corporate-wide consideration.

The software and hardware of GIS is only one component in GIS project implementation. It may be the biggest initial investment, but it is only a relatively small part of the total cost of a whole project. The data is the fuel for a GIS engine. Data not only consumes the biggest share of GIS long-term operational budget, but also causes most problems in any GIS project. GIS implementation not only will change certain processes in a system, but also will have significant impact on the organizational structure, financial commitments and cultural aspects.

On the other hand, since large-scale GIS project implementation is long term, there is a high risk associated with the process, changing organization objectives and advancing in technology. Replacement of decision makers and staff is involved. Also, revising financial plans can always bring significant risk to the successful GIS project implementation.

The 1980s has been a time of learning, experimenting and exploring in the implementation of geographic information systems. The cost paid is heavy and the experience gained is invaluable.

This chapter builds on experiences, and provides participants in an organization with guidelines for each phase of the implementation process. It outlines a reliable means for planners and decision makers to implement an effective GIS in their own organization.

In reality, political influence, personnel preferences and business culture are inevitably associated with GIS implementation. There is no effective way to exclude such influences from a systematic and rational approach, and it is not the intention of these guidelines to encourage the planners and decision makers to put all these issues above a normal scientific approach. It is recommended that all planners and decision makers make good use of these variations, in addition to the procedures recommended to achieve the maximum benefits, in a broad aspect, and in the long term.

4.2 Methodology

The GIS acquisition framework is built on several project-planning and management methodologies. These include strategic planning, business re-engineering, information system development cycle, and general information technology acquisition methodology, together with computer-assisted project management (CAPM) tools.

4.2.1 Theory and tools
Strategic planning

A disciplined effort is required to produce fundamental decisions and actions that shape the nature and direction of an organization's activities within legal bounds. This consists of the following major steps (Bryson 1993):

1. *Development of an initial agreement concerning the strategic planning effort.* It should cover:
   (a) the purpose of the effort
   (b) preferred steps in the process
   (c) the form and timing of reports
   (d) the role functions and membership of both the strategic planning coordinating committee and the strategic planning team
   (e) commitment of necessary resources to proceed with the effort
2. *Identification and clarification of mandates.* The purpose of this step is to identify and clarify the externally imposed formal and informal mandates placed on the organization.
3. *Development and clarification of mission and values.* This requires the development and clarification of the organization's mission and values. Prior to the development of a mission statement, an organization should complete a stakeholder analysis. A *stakeholder* is defined as any person, group or organization that can place a claim on an organization's attention, resources or output, or is affected by that output.
4. *External environmental assessment.* This is the exploration of the environment outside the organization in order to identify the opportunities, and threats that the organization faces.
5. *Internal environmental assessment.* This is an assessment of the organization itself in order to identify its strengths and weaknesses. Three assessment categories include organizational resources (inputs), present strategy (process) and performance (outputs).
6. *Strategic issue identification.* Strategic issues are:
   (a) fundamental policy questions affecting the organization's mandates

   (b) mission and values
   (c) product or service level and mix of clients, users, cost financing, management or organizational design
7. *Strategy development.* Strategies are developed to deal with the strategy issues. A *strategy* is a pattern of purposes, policies, programs, actions, decisions and/or resource allocations that define what an organization is, what it does and why it does it.

A GIS project in developing countries, especially in the public sector is also a political issue. Strategic planning could be used to identify strategic issues, and develop a *GIS acquisition strategic plan.*

System development life cycle

The *systems development life cycle* (SDLC) is an organized approach used in the organization to develop an information system (Shelly 1991). The systems development project life cycle consists of the five phases shown in Figure 4.2.

This is a standard procedure for conducting *information technology project management.* It could be used in GIS implementation at the tactical level.

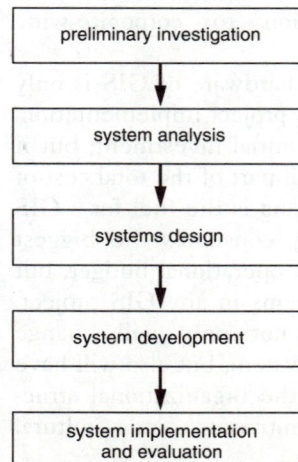

Fig. 4.2 *Project life cycle*

Acquisition of information technology– conceptual framework

The framework is part of a conceptual model relating to the acquisition, implementation and operation of 'information technology based systems' in any organization in government or major institutions (NSW Government 1991). Such a conceptual model:

- recognizes that the entire information technology acquisition process is driven by the business needs and direction of an organization
- reflects the cyclical nature of the process relating to information technology planning, implementation and operations
- illustrates the progressive or sequential relationship between strategic and tactical planning, and also between tactical planning and selection.

The framework contains the five successive steps shown in Figure 4.3. This approach could be used to address some of the GIS acquisition issues.

Computer assisted project management software

Project management is the process of planning and the management of time, equipment, cost,

Fig. 4.3 *IT acquisition framework*

people and tasks. To a GIS project, the geographic data could be another variable. CAPM software has been available since the 1980s, when it was downgraded from a professional's tool to an ease-of-use application software, along with word processors, spreadsheets and graphics. All projects are divided into two main phases: planning and implementation. Broadly, the central purpose of *project management software* (Behrsin, Mason & Sharpe 1994) is to:

- enable one to design, build and document a plan or model of how the project is expected to evolve over time
- use the plan to monitor progress during implementation in helping to ensure that the original schedules and budgets are met

The use of such software encourages a streamlined and systematic approach towards the process of project planning. It is a very useful tool in the whole of GIS project implementation, GIS technology acquisition and operation.

4.2.2 GIS technology acquisition management framework

The methodologies mentioned above all are very useful in managing GIS implementation. Since the need for application of a GIS is an important issue in many public and private organizations, the strategic planning and business engineering can be used to define the project at strategic level. Technologically, a GIS is a technology between a 'turnkey' technology and a brand new information system that would need to be developed, including the system development life cycle and the framework for acquisition information technology. Both could be used in managing GIS implementation.

The framework consists of four major stages (Fig. 4.4):

1. *Prefeasibility investigation*—using strategic planning to address some of the most important issues at the strategy level and build a strategic GIS implementation plan

2. *Feasibility study*—using the framework for acquisition of information technology, and addressing some other issues relating to data sources, data conversion approach, organizational structure and human resources
3. *System selection*—using the framework for acquisition of information technology, and addressing other issues, such as the standard and system operational environment
4. *System development*—using the system development life cycle to make the GIS workable within an organizational environment

The advantage of this framework is to reduce the risk of failure to the minimum. Each stage is a smaller cycle within the big project. At each stage, the applicability of GIS technology, the cost/benefit indication and the impact on the organization are assessed. The process at each stage involves understanding the organization's aims, functions and problems, and reducing the associated uncertainty. Gain for the organization is usually proportional to the resources invested. The organization itself (not the project and not the vendors) has total control over the project, that is, it can stop or proceed with the project when any strategic or tactical issues emerge.

4.3 Prefeasibility investigation

The initial stage of the prefeasibility investigation will involve:

- familiarization with the operations and long-term plan of the organization
- understanding of the current and target information technology strategy
- identification of strategic opportunities for using a GIS to better support the target environment
- preparation of a draft project plan and, possibly, a detailed action plan for the feasibility study, if the recommendation is positive

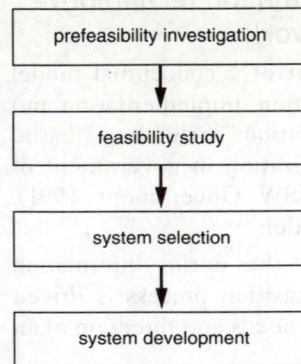

Fig. 4.4 *GIS implementation framework*

To start an investigation, a formal or informal *terms of reference* (TOR) should be prepared as a starting point. It should include the background information, the objective of the study, the methodology that should be used in the study, the financial and human resource allocation, a time frame and guidelines for reporting.

The concept of acquiring GIS technology and TOR are often introduced by the GIS facilitators. However, such an investigation is conducted by a committee consisting of a senior executive, who will chair the investigation team, GIS facilitators and/or external consultants. Well-balanced administrative power, internal enthusiasm and experience, and external expertise, are the key to a fair and successful prefeasibility investigation.

4.3.1 Reviewing a range of the organization's objectives

Reviewing the organization's objectives seeks to understand the business operation and the functionality of the organization. It is also a process to find the link between objectives and the information requirements at the strategic level. It is necessary to review the short-term plan as a means of assessing the existing systems. The long-term plan will indicate the organization's projection of operational activity in the future. Every organization has its long-term goals or aims for the organization. As an example, local governments usually have a long-range planning function, such as a 3-year plan, guided by the *Local Government Act* and other laws, which outlines the direction in which the prevailing administration wants the local governments to follow. Such long-range plans are quite general in nature, focusing on trends and problems requiring long-term programming, such as zoning, job creation, improving the economy, environmental improvement, maintaining infrastructure, improving the quality of neighborhoods, and a host of other local issues that affect the local residents.

4.3.2 Evaluating the geographic information needs and processing in the business environment

Since information is also a resource, it is very important to have an information technology strategy to support the organization's objective and operation. The need for a strategy for adopting information technology is one of the priorities in most organizations in the world.

It is necessary to prove that geographic information technology is the priority in an IT strategy. The advantage of a GIS is the capability to process the geographic information. This process is to identify the proportion of geographic information in a business environment. For example, geographic information in a mapping authority and utility industry is close to 100%, in local government 70–80%, and an environment protection agency perhaps 80%, and so on.

Another important step is to investigate the geographic information processing system, whether it is manual mapping or using computer-aided mapping (CAM), manual processing of textual data or using a computerized database management system (DBMS), integrated published maps and tables or separated maps and documents, and so on.

The next step is to find the problems of a current geographic information processing system. Although it is one of the driving forces

to carry out this project, a prioritized *problem list* should be produced, such as:

- how to manage the vast amount of geographic information
- how to overcome the difficulty of keeping the geographic information updated/upgraded with speed
- how to overcome the long time to do map search and map analysis
- how to overcome the difficulty to produce customized geographic information
- how to overcome the difficulty to conduct a search between graphic data and textual data
- how to ensure that the quality of current products can meet the planning, management requirements, and so on

A GIS with the capability of data integration, data analysis and digital mapping is a possible solution to the above problems. But there are some other solutions, such as a *relational database management system* (RDBMS), a digital mapping system, a computer-aided engineering drafting system, and visualization, animation and simulation systems. It is important to prove at the initial strategy stage that a GIS is the best solution to the problems and that it is a strategic issue.

4.3.3 Formulating a long-term GIS project plan and alternative strategies

After a GIS is identified as a prioritized solution, an inclusive long-range plan, which investigates the needs of the organization over a long period of time, should be compiled. This ensures that the use of the GIS technology will agree with the tactical and strategic goals of the organization, and the strategy for the use of the information technology in the organization. This reduces the risk to investment, unrealized expectations and possible disappointment in the system, as has happened repeatedly over the past 20 years.

The long-range plan allows decision makers to evaluate the applicability of the system as it is relevant to the strategic and tactical plans of the organization. Also, it ensures that the appropriate financial and human resources, as required, are available at the time they are needed, that is, during the process of implementation and further into the operation.

If a long-range plan is successful at identifying all potential applications, and is related to the strategic and tactical plans of the organization as a whole, then it is possible to schedule or prioritize the implementation of applications in such a way as to obtain most benefit to the organization. It also ensures that those applications or databases that are needed to be in place are actually in place as scheduled, to ensure continuing development or expansion. The strategic plan will decide if the GIS project is sustainable, while the tactical plan will decide when the GIS project should start.

The alternative GIS implementation strategies should be designed with the degree of restructuring in an organization, and identify separately each major GIS resource component, that is, data, applications, technology and organization. Using GIS technology is not only the automation of existing manual methods, but also the revolution of work procedures and, further, the organization's ability to deliver products. The analysis should focus on:

- optional strategies that are available
- the advantage and disadvantage between specific strategies
- impact on the organization of adopting the new strategies
- known resource or funding constraints
- all costs associated with the acquisition, the implementation and the ongoing maintenance

It is also very important to obtain *executive support*. The long-range plan not only brings opportunities for improving the organization to the attention of decision makers, but also gives them confidence that those who advocate the new GIS technology are competent and can make the project succeed.

4.3.4 Selecting preferred GIS strategy

The time will come for the selection of one preferred GIS strategy, and the development of an *action plan* that identifies the nature, cost and timing for the implementation of the strategies. This selection will need the approval of senior management or committee of the organization. The report should include:

- the history and the background of the project
- description of the strategy proposals
- assessment of selected strategy
- project plan
- terms of reference for the implementation of the feasibility study

4.4 Feasibility study

The focus of the feasibility study for the introduction of a GIS into an organization is to undertake a *user requirement study and analysis* to identify all issues of the organization that can benefit from an appropriate GIS specification. This can provide the implementation framework for such a system.

4.4.1 Project preparation

If the feasibility study has been approved by senior management or committee, it is the time to start the project preparation.

Organizational issues

The organizational procedure for a GIS project is outlined in Figure 4.5.

Executive steering committee

As a means of ensuring that all departmental functions are properly represented in the designated implementation of a GIS, it is always advisable to create an executive steering committee, comprising the senior executive officer and higher ranked officers from the relevant departments, for example, managerial, technical, accountants and professional departments. This coordination between all executives will facilitate the GIS project implementation.

```
       ┌─────────────────────────────┐
       │ executive steering committee │
       └─────────────────────────────┘
                      │
                      ▼
       ┌─────────────────────────────┐
       │    technical committee       │
       └─────────────────────────────┘
                      │
                      ▼
       ⬡ GIS project coordinator ⬡
                      │
        ┌─────────────┴─────────────┐
        ▼                           ▼
 ┌──────────────┐          ┌──────────────────┐
 │  URAL team   │─────────▶│  cost – benefit  │
 │              │          │  analysis team   │
 └──────────────┘          └──────────────────┘
        │         ╲                  │
        ▼          ╲                 ▼
 ┌──────────────┐   ╲       ┌──────────────────┐
 │ training need │   ─────▶ │  risk analysis   │
 │ analysis team │          │      team        │
 └──────────────┘          └──────────────────┘
```

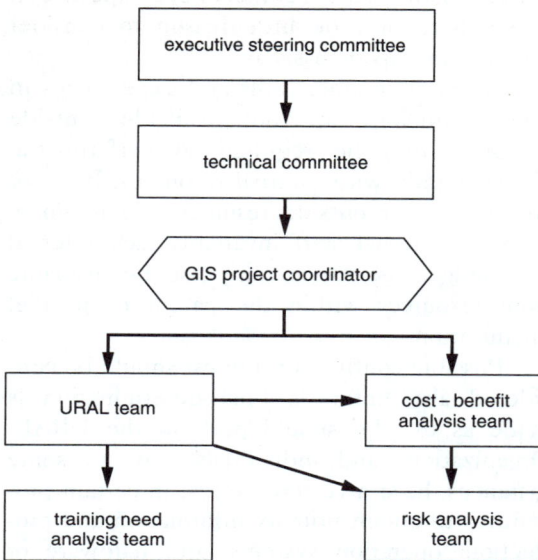

Fig. 4.5 *Organization structure for a GIS project*

The functions of the executive committee, at all steps, should include:

- supervision and implementation of the project plan
- decision making based on the technical committee's recommendations, regarding funds, human resources, equipment/infrastructure, and so on
- liaison with other related organizations

The executive committee usually organizes a technical committee to ensure that the feasibility study is undertaken in the appropriate manner.

Technical committee

The technical committee usually comprises a GIS coordinator, the manager of information technology and the managers, or most senior members, of each of the technical sections, such as planning, engineering, surveying, map production, utilities, finance, training, and so on.

The functions of the technical committee are to:

- ensure effective implementation of the project plan

- organize and supervise user requirement analysis, cost/benefit study and risk analysis appraisal
- review all reports and documents as well as specifications and bid documents, including:
  - specifications for the scope of work for an external consultant, if required
  - primary data specifications
  - conceptual design
  - detailed system design
  - hardware and software requirements
  - data conversion
  - pilot projects
- identify and resolve the detailed coordination issues that are not covered in the strategic plan, including, but not limited to, data, standards and custodial responsibilities
- provide background and support information to the executive committee
- update the project plan
- recommend executive action

Seminar to introduce the GIS project

It is advisable to prepare a GIS seminar to introduce the GIS project to all departments. The purpose of the seminar is to introduce members of the executive steering committee, the technical committees and the GIS consultant to all personnel of the organization who could be involved. The seminar could be a half-day or a full-day presentation. The session should include a series of remarks by department representatives who present the project, the goals of the project, and the people who would be involved. This could be followed by a brief introduction to GIS technology and terminology, and an industry overview, with an outline of the current and proposed future activities.

A handout package should be created, which would provide every participant with reference material, in a planned endeavor to provide continuing project documentation, education and updating of personnel on the GIS project. This initial package could contain:

- GIS brief
- project history and overview
- methodologies used in the project
- deliverables in the project
- project organization chart
- project activities chart (task, time, period, characteristics)
- possible software and equipment configuration
- assistance required from different departments
- glossary of GIS terms

4.4.2 User requirements analysis

A user requirements analysis (URAL) is a comprehensive study of the needs of all types of proposed GIS users such as *system users*, *end users* and *potential users*, but is not limited to the study of the people only. The URAL should result in a clear statement of primary, interim and end product characteristics, required production rates, and estimate data volumes, and should provide a base for cost-benefit analysis, along with a system conceptual prototype.

URAL team

A URAL should be performed by in-house staff, or through a combination of *external consultants*, and *in-house staff*, who should report to the technical committee. Through the URAL, the requirements of the users can then be matched with system capability to determine optimal configurations for the organization's GIS procurement.

The important element in determining who should perform the URAL is to select those with a thorough understanding of both GIS technology and the operations of the organization. It would be the responsibility of the technical committee to assure that the URAL fully understands the organizations's objectives, plans, products, services, mission and needs.

In-house staff naturally have a greater understanding of the tasks and the procedures for the system that are to be considered for computerizing through GIS technology. In cases

where existing staff members have expertise in GISs, there may be little reason to consider bringing in outside assistance.

When staff time, skills and experiences in GIS technology are not available, outside resources may be required to perform the URAL jointly with internal resources. It is not suggested that outside resources do it alone, because external staff invariably lack internal knowledge. Assistance may also be available from resources within the parent or parallel organization.

Possible conflicts of interest should be considered also before a final determination is made as to who should perform the URAL. Organizations and individuals may, in some instances, have a vested interest in certain procedures to obtain primary information and production, operation systems, and hardware or software types, and may be inclined, whether intentionally or unintentionally, to bias results of the URAL towards particular solutions and systems. The objective of the URAL is to identify the needs of an organization and then to select the system, if such a system exists, that best fills those needs. All reasonable effort must be made to assure this goal is realized, including assessing possible conflicts of interests or biases on the part of persons or organizations that potentially could perform the URAL.

Methodologies to conduct URAL
Interviews

The interview is an important tool for conducting a user requirement analysis. It is recommended that it should be carried out before the questionnaire survey, because it could help the questionnaire to be more precise.

The individuals to be interviewed are derived from numerous sources. Selection should be based on position title, and recommendations made by department heads, who may provide names of appropriate individuals within their departments. Additional interviewees may be obtained during the interview process.

The majority of interviews should be conducted on an individual basis, except when a department requests a group interview. All interviews should be conducted on a team basis, ensuring continuity along departmental lines and areas of organizational services.

A pilot project, utilizing different approaches, could be undertaken with a few initial interviews prior to finalizing the interview format.

Transcripts should be returned to the interviewees for verification of the information that was obtained during the interview.

Departmental heads need to be given a printed copy of the summarized interview notes in draft form, and to make appropriate comments before the final release of the user requirement analysis report. A minimum of comments should be received and incorporated into the final user requirement analysis report.

Questionnaire survey

A survey questionnaire should be designed to provide details about all types of *users, primary information, processes, interim and final products, applications, attitudes and potential GIS-related activities* within the organization.

Recipients of the survey questionnaire should be selected organization-wide and, if relevant, this should be extended to some related outside agencies, such as parent and parallel organisations, clients, contractors and partners. The selection process should consist of the review of all position titles within the organization to determine those with a high potential for utilization of a GIS system. In addition, the survey questionnaires should be sent to all remaining employees who may become potential users.

The questionnaire should be designed with the combination of predefined questions and a few descriptive questions. Much depends on the knowledge of the questionnaire design staff on the process of legislative aspects, management and GIS technology. The more time spent on the questionnaire design, the less time will be necessary on the analysis.

The results of the survey should be analyzed as soon as possible, and a summary of the results should be printed and sent to the persons surveyed and anyone else involved.

An investigation of relevant information

Relevant information may be the source maps, documents and information-processing logbook, process descriptions, quality assurance procedures and other records kept in the organization.

There are many conferences, workshops and seminars on geographic information systems and applications held around the world each year, and numerous publications on GIS applications. With an *inclusive research into applications* in the comparable organizations, the results can provide the direction of GIS applications, the general user requirements, experiences, failures and pitfalls. This research will assist questionnaire design and critical issues selection for interview, reduce the work load on the survey and interview significantly, and improve the confidence of a user requirement research team.

Identifying users

Identifying the users is to pinpoint the current system users, draw the picture of future users, and measure the difference between them. This will form the sample pool for the interview and questionnaire survey.

A *GIS user*, in general, can be defined as a person who works with GIS products developed by the system or who uses GIS for planning, decision making, management and operation. The *potential user* is someone who cannot use the present system because of some constraint, but could become a user of the system if it were converted to a GIS.

Interviewing senior executive officers, departmental heads and key professionals will assist in defining current users and the proposed GIS users. Further, the current users will be the sample pool for a questionnaire survey.

The information collected for identifying users could be used for a further *staff training needs analysis.*

Current system and geographic information products analysis

After the investigation of products and information required from within the organization, it is necessary to study the current information processing system. There are three aspects in terms of current system analysis: current manual system analysis, current digital information system analysis, and current CAD/CAM system analysis.

Current manual system analysis

The aim of conducting a current manual system analysis is to provide the basis for GIS applications. Beginning with the *function analysis* over the organization, all functions that contribute to the goals of the organization must be reviewed to determine how a GIS can be used to assist in their improvement. These functions or tasks are usually defined in the mission statements of the various departments, bureaus, divisions, sections or other offices within the organizational structure.

Detailed information on the current paper-based manual system should be gathered by a URAL team through interviewing the personnel involved with the existing system, such as the managers, professionals and technicians. The interviews should be supplemented by observation of their work, noting significant backlogs if any, and the applications of various data sets that are made by the staff. Particular care must be taken to identify the specific types of data used, including the data topology, format, media, representation and accuracy. At the same time, documentation concerning the costs of operating the existing system should be developed. The results of the review of the existing system will serve as the basis for cost-benefit analysis.

Current digital information system analysis

The objective is to find the compatibility between the proposed GIS and existing digital information system.

Many organizations have already implemented digital information systems on existing computers to support their nongraphics data needs (word processing, financial management, register, inventory, and so on). Some of these systems may have been operating successfully for many years. Existing digital information systems may contain nongraphics attribute data that are important for use in a GIS. The *data needs interviews* with people who use these nongraphic data systems can provide valuable information for determining how a GIS should be implemented and integrated with these existing systems. If many problems are experienced with an existing system, it may be beneficial to redesign it on the GIS. However, if an existing system provides adequate information service to its users, an interface between it and the new GIS may be considered more beneficial, thereby avoiding the expense of redesigning it for the GIS. Interfacing a GIS with an existing system can be achieved either by transferring files to the GIS, or by processing transactions from the GIS directly to the data base of another system and then back again to the GIS.

If appropriate communications hardware and software are installed so that the GIS can be connected to a multisystem network, then direct file transfers (moving entire databases from an existing system to the GIS) are possible, as are direct database accesses between systems. These existing information systems should be studies in depth during the *geographic information needs study* in order to evaluate their value and determine whether they should be interfaced to the GIS, abandoned or redesigned on the GIS.

Current CAD/CAM system analysis

Computer-aided drafting/mapping systems revolutionized the way of mapping in the early 1980s. Many organizations have a CAD/CAM system to meet the requirements of the graphic representation of geographic information. Also many sophisticated software packages have evolved from CAD/CAM, and there are many GIS modules running in a CAD/CAM environment. If your organization has a CAD/CAM system, the following issues related to the transformation of CAD/CAM into a GIS should be analyzed:

- software specification
  - data capture capabilities
  - mapping capabilities
  - availability of analytical modules
  - development environment and tools
- CAD/CAM data
  - drawing data specifications, including quantity, quality and format
  - spatial data exchange with further GIS
  - associated attribute data availability
- equipment for CAD/CAM
  - computer
  - network
  - digitizer/scanner
  - plotter
- organizational issues in CAD/CAM
  - executive attitude
  - human resources and expertise
  - working spaces and furniture
  - financial resources

Some of the sophisticated CAD/CAM software have excellent spatial data edition and mapping functions. CAD/CAM could be considered as the alternative or supplementary tools for data capture and digital mapping. It is worth investigating the availability of GIS modules built in the particular CAD/CAM software.

Identifying GIS applications

The GIS function must be matched to the need of geographic data collection, processing, manipulation, analysis and presentation. You can use the GIS function illustrated in Chapter 3, pages 52–62. This process will later be transformed into the GIS system tendering specification for software/hardware vendors.

Refinement of the kinds of GIS products and services

There are some differences between the GIS products and existing geographic information products. GIS products may be the hardcopy on paper; display on the monitor; projection on the screen; digital data with various topological relationships; and information on the diskette, type, CD, and so on. The initial definition of required products and the evaluation of the current system provides a description of products currently being provided to users. However, the characteristics of those products should be thoughtfully reassessed before they are used as the criteria for selecting a GIS. The final product definitions should reflect the flexibility of the current GIS technology in generating products, and meeting various needs of user groups.

In many cases, products from the existing system have been developed to serve a large number of multidisciplinary users. As such, these products may contain a large amount of information irrelevant to the applications of an individual user. Or an individual user may find a need to reformat, rescale or extract information from the existing product in some manner before this data can be applied to the user's application. Analytical procedures, even those within a single discipline and performed within a single office, may differ drastically in required end products. Often tabular data is produced from GIS-based analysis, while in other instances high-quality graphics will be required. Frequently, a combination of data presentations will be required. In defining GIS products, the flexibility of the output from a GIS should be evaluated for providing *tailored products* specific to *user applications*. The URAL team must be aware of the full range of desired end-product presentations.

Issues relating to the formats, media, representations and accuracy of the data required by the user must be carried over into the consideration of GIS end-product characteristics. GIS *output capabilities* should be designed to provide the product that the users require to perform their missions. GIS products may have a direct input to models, incorporated within digital data sets, or stored on computer-compatible media. Users requiring these GIS end products may need digital data formats and information types. In the design of the GIS, the input formats and data requirements of user computer models should be reviewed.

A definition of GIS product characteristics should result in a clear statement of:

- source document information (scale, accuracy, projection, version, and so on)
- presentation media
- accuracy
- information representation
- information format
- information content
- digital information types and structures
- analytical model interfaces
- the rate of the production

Refinement of the amounts of GIS production and services

Production rates for the GIS must be specified so that the system can be designed to meet organizational objectives. If the GIS is intended to develop a well-defined product for mass production, the production rate is simply that stated for the organization. The production rate in terms of digital data can be calculated from an estimate of the digital data volume of a sample of products.

For a GIS that is intended to *support analytical needs* for planning and decision making, the production rate is difficult to quantify. Studies that will rely on a GIS to analyze spatial data are not normally defined in terms of GIS products required to support the study. The production of these analytical products of a GIS is limited by the ability of the system to retrieve, analyze and display information from the database. The number of users expected to be accessing a system simultaneously, along with the number accessing the data processing requirements, must be estimated to determine GIS configurations that are capable of providing the *level of user analytical support* desired by the organization.

Estimates of required analytical products can be made by reviewing past studies by the organization to estimate the number of final and draft map images, graphics, statistical analyses and other GIS products required. Interviews with the staff can identify additional products that might have been produced if GIS capabilities were available. Based on this past experience, the GIS production rates for analytical studies can be extrapolated.

Production rates have the following five primary constraints:

- availability of digital data
- personnel's operational and analytical skill
- availability of hardware and peripherals
- the performance of hardware and peripherals
- the management

The number, types and purpose of products required from the GIS will refine actual system configuration.

Conceptual database design

The conceptual GIS database design provides a logical view of specific applications and the data relationships that are necessary to enable these applications. It will investigate the database model, the data structure of spatial data and attributes, and the data sources for the database structure. The conceptual design should demonstrate compliance with the corporate view of data, and its integration with other corporate databases, in addition to the maintenance and support needs. It is a prerequisite to physical database design.

Source data

In case some of the current analog and digital data are not available, or the accuracy, currency or other requirements do not meet the data requirements, the URAL team should also address the source data acquisition issues at this stage. The methodology used to obtain source data should be determined. Current available technologies are satellite remote sensing, aerial photogrammetry, small-format aerial photogrammetry, field survey, GPS survey, and so on. The selection criteria are data quality, data requirement, cost and availability of technology.

Remote sensing

Remote sensing data covering the whole world is now available. If the GIS software can accommodate *processed images in raster*, direct use can be made of the data for operation and analysis. The image extraction functions are also provided by some well-developed GIS software.

Aerial photogrammetry

Aerial photographs and photogrammetry have been a highly developed and accurate system for supplying survey and map data for nearly 80 years. This system has been used mostly for national topographical and environmental mapping at a scale of 1:50 000. It has become comparatively costly and slow to deliver some maps such as large-scale town maps.

Small-format aerial photogrammetry

In recent years, the demand for large-scale (1: 1 000) accurate town maps at low cost has been met by an Australian company, Advanced Design and Manufacturing Co. (ADAM). This company has produced an analytical stereo plotter that, using the 56 millimetre × 56 millimetre (2.2 inch) diapositive (small-format camera diapositive), can provide the required town maps with the same degree of cadastral accuracy as that provided by the 23 centimetre (9 inch) metric camera.

Field survey

The digital data recorder is generally available with the *modern total station*. Data can normally be available in ASCII. It is necessary that a GIS support GOCO (Geometry Coordinate) be included, which can communicate with CAD software, such as AutoCAD or Microstation. Global positioning system (GPS) receivers can be used to obtain the necessary control and other data.

Data-capture options

The URAL team should decide the portions of *in-house capture* and *contract capture* at this stage, because this directly affects the sectoral allocation of financial resources and other organizational issues. The major factors that determine the 'best' option can be considered to be those of initial cost, program for completion, availability of in-house resources, culture of the originating organization and the timing of the financial return demanded.

In-house GIS data capture

In-house GIS data capture has proven to be very successful in those organizations that need to maintain control and long-term investment in their GIS, or if a reliable independent GIS data-capture organization is not available locally.

The GIS investment can be made more effective if the resources purchased are used on a round-the-clock shift basis. This can be achieved by employing and training additional temporary staff for the conversion phase of the project. The extent of training required should not be underestimated.

Data capture, undertaken internally, maintains total control of the flow of the *conversion project* and permits quality control to be attained without the problems arising from data transportation between different computer systems. Those most closely involved in the records on a day-to-day basis would be immediately available for resolution of errors/ omissions.

Output is constrained by the volume of equipment purchased, space considerations, and by the need to recruit and train temporary staff.

The use of in-house data capture necessitates that supervisory staff be willing to operate unsocial hours on a continual basis during the data-capture phase.

The organization in which this task is undertaken could consider deploying a suitable reward system that encourages high productivity of quality output.

External contractors

Experience from developed countries indicates a preference for external contractors. After considering the resources (financial, human resources and equipment) and schedule demands, many organizations, and most organizations which are just starting a GIS project, lack the knowledge and in-house expertise, trained personnel and equipment, and special conversion software, to perform capture in-house satisfactorily.

The speed of data capture can be expedited by the use of experienced external data-capture contractors.

Where careful selection of external contractors has been undertaken, many of the preferred companies have established their own rigorous in-house quality assurance procedures, which will meet the most demanding of standards (where these have been fully defined within the specification at the outset of the contract).

The data-capture industry may provide a very competitively priced service. These companies have a very real understanding of the data-capture process, as their success or failure depends upon it, and they will select the correct method for the efficient running of the project.

Data-capture companies employ experienced engineers from the very diverse range of industries with whom they are expected to deal, for example water, electricity, telecommunication and gas.

Budgets and schedules that may be hard to meet or control internally become the responsibility of the *conversion service company* through contractual arrangements. Capture companies rarely underestimate the cost, whereas in-house projects often do not include sufficient resources, or do not provide for the hidden costs, rigorous quality control and acceptance procedures.

Combination of in-house and external contractors

The installation phase of most GIS projects is often the most difficult data-capture period. During this period many issues emerge for the first time. The data model, which forms the foundation of the GIS, needs to be tested rigorously. This cannot be achieved without the availability of data. The placement of commercial data-capture contracts at this stage is a risky strategy.

The in-house team will be fully aware of the data problems that are being experienced by external contractors. In addition, it provides valuable feedback on the costs being incurred on the organization's behalf, and creates a useful comparison for future conversion projects.

Small-scale in-house capture is absolutely necessary in terms of making full use of existing equipment and human resources. On-job training of staff for database updating and upgrading, and on-job training for end users, are necessary to familiarize themselves with the system, and for improving the communication with the data-capture companies.

4.4.3 Training issues

Experience has shown that effective staff training is one of the critical factors to the successful implementation of a GIS project. In the past, hundreds of millions of dollars invested in GIS facilities have not been cost-effective, as the GIS do not operate adequately because of insufficient and/or inappropriate training. This matter is addressed separately in order to generate an adequate awareness of the proper training requirement.

Training needs analysis and staffing

Each GIS application is new to most users, making a training program essential. In any situation where there is a gap between present job performance, and desired job performance, a *needs analysis* can be used to identify the best solution. Needs analysis techniques are basically the same, whether one is evaluating GIS operations or any other technical task.

Identifying the *information needed* to create a new GIS training program requires that the GIS/training coordinator should know who the users are, what their tasks entail, the type of data they use, and how they will use the GIS. In short, the training officer must know detailed information about the user profile, common procedures, typical data and typical scenarios before an effective program can be developed. Much information can be found in primary information from a URAL and its conclusions, as well as from appropriate educational/training institutions.

There are several ways to compile or summarize the data that is retrieved through the needs analysis. *Training-needs matrices* can clarify much of the subject matter on which the specific training is required.

The information must be compiled into a *training requirements document*. This document is the foundation on which the training program rests. Since the software and equipment have not been selected, only a draft framework for a training program should be prepared. The training program should be finalized before the operation of a GIS.

The framework for training

Since the training issues involve many other factors, such as the function of various personnel, subjects of training, stage of training, who delivers the training and how to deliver the training, there should be a framework to address these issues, including the following:

- Personnel functions are:
  - manager
  - professional staff
  - database operator
  - staff for data capture
  - end user
- Subjects include:
  - background knowledge: mathematics; information technology; survey and mapping; social economy; engineering; and so on
  - core GIS: GIS introduction; data capture; database modeling; database management; analysis; digital mapping; applications development; production management; quality assurance; and so on
- Timing of training involves:
  - feasibility study

- system selection
- data capture
- database manipulation
- GIS operation, and so on
- Training institutions include:
  - university
  - GIS vendor
  - consultant
  - third party institution
  - internal, and so on
- Delivery arrangements include:
  - classroom
  - hands on
  - study tour, and so on
- Location involves:
  - in-house
  - institution
  - industry

As an example of the structuring of a training program, the Australian Institute of Spatial Information Sciences and Technology (AISIST) has developed an outline of short training modules to facilitate GIS training in key management and operational areas. The program comes together through a hierarchical structure (Fig. 4.6) with participants working through a structured program of 'competencies', from chief executive officer to management levels, and supervisory levels to operator skills. At the lower levels the generic modules are linked to vendor specific operator training. The structure was compiled by the co-chairman of the Scientific Advisory Council of AISIST, Professor P. Zwartz, Centre for Spatial Information Studies, University of Tasmania.

This hierarchical design incorporates a number of educational features:

1. Levels 1-3 are specifically designed to provide the broad, conceptual background considered as essential prerequisite knowledge for all staff, before proceeding to the more practical Levels 4 and 5. Staff and participants who later complete units at Levels 4–5 will have received the necessary theoretical underpinning to allow them to gain maximum benefit.

| Level 1 | GIS introduction | | |

| Level 2 | GIS data | | database issues |
| | spatial modules | | corporate issues |

| Level 3 | analysis | spatial representation | managing GIS introduction |

| Level 4 | analysis, modeling and output | | data sources, input and verification |
| | organizational issues | databases design | GIS manipulation |

| Level 5 | Genamap | vendor specific | ARC/INFO |
| | Mapinfo Intergraph | | Geodis |

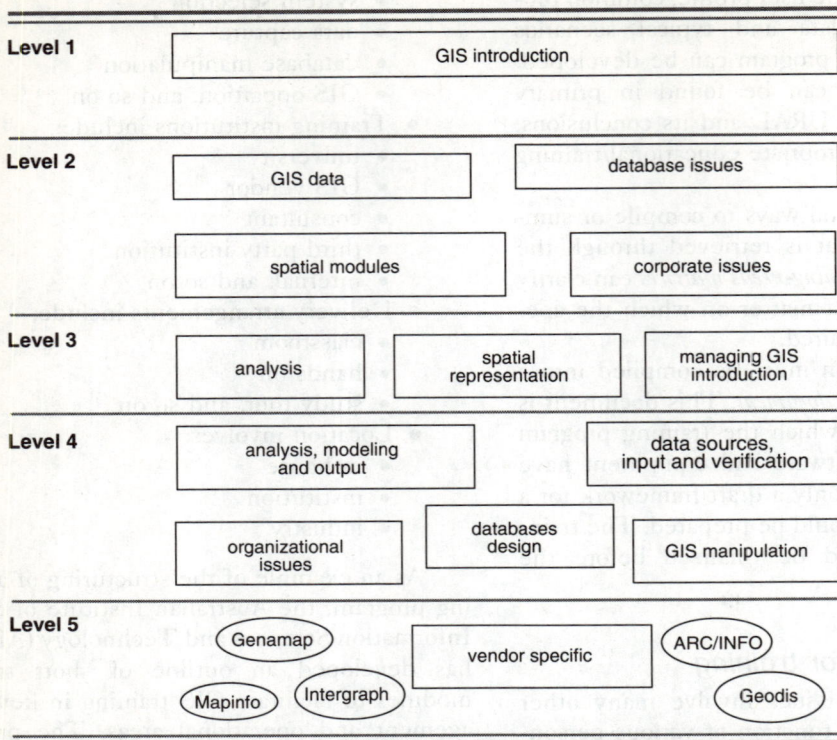

Fig. 4.6 *AISIST training modules structure*

2. Each 'study level' targets personnel at different occupational levels in an organization. For example, Levels 1 and 2 target managers, and Levels 2 and 3 target senior technical specialists and section managers.

3. Normally all students will commence at Level 1, and proceed down the hierarchy, but only as far as their particular task requires the specialized knowledge.

4. There is a steady graduation from the theoretical levels (Levels 1 and 2) required by managers, through to the practical levels (Levels 4 and 5) required by operators.

5. In the same way as there is a carefully structured vertical development of conceptual matter from Level 1 through to Level 5, there is also a deliberate attempt to develop horizontal integration. For example, all units offered at Level 3 require a similar level of conceptual demand. Across Level 3 further integration is apparent within modules.

6. The curriculum is designed to be flexible, and to cater for the different needs of clients. One person may require a broad introduction to the whole area and can achieve this by completing all the units in Levels 1–3. Another client may identify a specific training need at Level 5, but first will complete the required prerequisite units at Levels 4 and above. In this way, 'pathways' can be recommended. Students are, in effect, able to dip-in where and when the training is appropriate, and benefit is gained by following a predetermined pathway

through the hierarchical structure. The system is flexible enough to allow industry practitioners to select individual modules dependent on their levels of expertise or their current needs.

Staffing

As illustrated in Chapter 2, several kinds of human resources are involved in a GIS project. Most of the positions could be filled by existing staff with the appropriate training. Hiring some external specialist, and an experienced person, is a good option to consider when the expertise cannot be found in the organization. If data capture is undertaken in-house, a number of temporary staff may be required.

4.4.4 Cost–benefit analysis

A cost–benefit analysis is a technique for systematically estimating or reviewing the efficiency impacts of policies or programs. A *cost-–benefit analysis* attempts to determine the costs of implementing a GIS and to estimate the benefits that GIS implementation would provide compared with the present system. It is used to justify (or even reject) a policy initiative to implement a GIS. The major information sources for cost–benefit analysis are from a user requirement study (URAL).

A cost–benefit analysis requires listing, categorizing and monetarizing impacts so that the economic efficiency of policies or programs can be objectively evaluated. Although all GIS implementations encounter different costs and benefits, there are many similarities. Typical cost and benefit factors found in a GIS include, but are not limited to, the following:

Cost

- Capital costs are:
  - prefeasibility investigation
  - feasibility studies
  - hardware
  - software
  - source data acquisition
  - possible digital data purchase
  - working environment upgrade
  - supporting equipment acquisition
- Personnel costs are:
  - consulting services
  - project management
  - database conversion
  - database maintenance
  - training
  - new position for systems, technical and operations support
- Operations costs are:
  - quality related
  - supplies
  - maintenance contracts
  - overheads
- Hidden costs are:
  - technology upgrade
  - transaction in organization
  - cost for staff reduction

Benefits

Benefits are:

- time and cost saving in information production/maintenance
- improvements in productivity
- improvements in data quality and validity
- quick and more versatile information processing
- better space utilization due to decreased physical storage requirements
- reduced data redundancy
- financial gains in terms of marketing new digital products
- better decision-making tools
- new analysis capabilities
- increased awareness and acceptance of higher quality standards
- increased compatibility and communication with other organizations
- increased corporate image
- increased public confidence, and so on

Implementation options

Detailed project options could be formulated based on the implementation options. There should be a time frame for each option, to arrange different tasks for primary data acquisition, software/hardware purchase, data capture and purchase, site preparation, financial commitments, training programs, and so on.

Cost–benefit ratio

The cost–benefit ratio is worked out to make an estimate of the tangible costs and benefits (cost saving, revenue) over the defined life of a GIS, year by year, in *present value* with an estimate *discount rate*. If the ratio of benefits to costs is much greater than one, the project is justified. The analysis should be undertaken over all draft implementation options, to rank each option in terms of financial advantage. If the ratio is close to one, further investigation over the intangible costs and benefits should be conducted to justify the cost.

4.4.5 Revising implementation plan

Only when the results of the URAL and the cost-benefit analysis show that there is no major negative impact on the GIS implementation, should the implementation plan be determined. This step will link the GIS implementation project with the organizations' objectives, long-term and short-term plan, financial management, human resources management and other arrangements. Such tasks involve:

- *Identifying tasks that need to be undertaken over the planning period*. These encompass data, applications, technology and organization strategies. The specifications will incorporate broad resource and funding estimates.
- *Finalizing project priorities over the planning period based on the criteria agreed with the steering committee*. These priorities will be linked to the projects to provide a sequence in order of importance to the organization. Any prerequisite or primary projects must be highlighted.
- *Developing a preliminary schedule, by major activity, for each project*. Where necessary, specific responsibilities should be allocated for projects and major activities. These schedules should be amalgamated

to form an overall implementation plan. Any appropriate methodology that will be applied must also be identified.

- *Converting the summary of the cost–benefit study into a financial plan, stating the assumptions, projections and associated risks*. This will include a statement of all quantifiable and nonquantifiable benefits associated with the plans over the useful life of the system.
- *Identifying the risks, if any*. These are the risks that will inhibit the successful outcome of the project, and the manner in which the agency plans to eliminate or minimize those risks.
- *Establishing the project timetable*. Identify the potential sources for the acquisition of equipment, software, technical and management support, education and training, primary data, available digital data purchase, and so on.

4.4.6 Preparing a formal proposal

The 'feasibility study team' can provide the technical committee with a comprehensive description of the impact on process, system, organization, finance, service and technology of a successful GIS implementation. Based on this information, the technical committee must evaluate the merits of GIS usage relative to the organization's *applications*.

The technical committee needs to review whether the URAL report is consistent with the goals and objectives of the organization, and determine if a system based on products from the proposed GIS technology will fulfill the organization's responsibilities.

The cost–benefit analysis is a very important indicator of the need for applying GIS technology within the organization. The degree to which intangible benefits have been adequately measured and quantified must be considered in a subjective manner when the cost–benefit ratio is evaluated. When the cost–benefit ratio is marginal, further research may be required, particularly into the intangible benefits, before a decision on GIS implementation can be made based on the cost–benefit ratio.

The technical committee should submit a formal proposal to the project steering committee for approval. The proposal may contain some of the following items:

- revised implementation plan
- draft specifications for data purchase/conversion, system selection, and training and system development
- supporting documentation: the feasibility study report, which would include the URAL report, the cost–benefit report and training framework, and the recommendations

4.5 System selection

The *system selection* is a process to select the appropriate software and hardware, but other issues, such as defining the work space, the human resources and data-capture options, should also be taken into consideration. The specification of the GIS software and hardware requirements in the feasibility study are reformatted for presentation to vendors. The hardware and software specifications should describe the functions required, and provide a general description of the hardware configuration. According to the specification, the implementation plan should be revised accordingly.

Generally, there are a few options that are not mutually exclusive to carry out this stage, depending on the scale and importance of the GIS project (NSW Government 1991), such as:

- Issue a *Request for Information*, based on an informal statement that asks for a capability statement of a company.
- Issue an *Expression of Interest*, based on an informal statement of the agency requirements. This may also be initiated to verify the nature, extent and cost of the available solutions.
- Issue a *Request for Proposal* (RFP), based on a more detailed statement of requirements than the Expression of Interest. It may be initiated to explore the range of

solutions that are available from the potential suppliers without binding either the agency or the supplier.

- Issue a *Request for Tender* (RFT), with the detailed requirements and conditions to elicit a comprehensive and comparable response from the suppliers.
- Use the *Period Contract Arrangements*, which are in force with respect to the GIS required.

If the GIS project is of strategic importance to the agency, the RFT-based process should be used. But the Expression of Interest and/or Request for Proposal may be used to facilitate the tender process. In this guideline the process of RFT is selected to illustrate this process.

The technical committee is responsible for the whole process. Internal staff and/or external consultants can be assigned to carry out the *system selection* task.

4.5.1 Revised URAL converted into system requirement

The *user requirement analysis* forms the foundations for system selection. However, the URAL describes the system in the language of the user, and cannot be issued to vendors as the basis for their tenders. A description of the required system in GIS terms is needed for the tendering process, and this must be written to reflect the requirements described in the URAL. It will be the selection criteria for the system selection. Following is a sample sublist that needs to be considered as far as the functionality and quality is concerned:

- software involves:
  - data capture (spatial and attribute data capture, editing functions, map projection, coordinate transformation and data import)
  - maintenance
  - management (internal database, database interface, flexibility, integrity, security, independence, recovery and backup, concurrence, performance, storage, and so on)

- manipulation and analysis (data query and spatial query, reclassification, dissolving and merging, buffer generation, polygon overlay and analysis, network analysis, proximity, and so on)
- output (display, plot, cartographic functions, reports and data exchange)
- interface (ease of use, screen and menu customization, microcommand, and so on)
- development environment (C, 4GL, and so on)
- standard (windows, open system, networking, and so on)
- Hardware involves:
  - operation system
  - CPU
  - memory
  - on-line storage
  - backup storage
  - graphic workstation
  - digitizers/scanners
  - graphic display terminals
  - alphanumeric display terminals
  - line/laser printer/verification plotter/electrostatic plotter, and so on
  - communication/networking facilities
  - special facilities

4.5.2 Preselection of suppliers

With several dozen fully-functionalized GIS suppliers in the world, it is neither desirable nor practical to go for a fully open tender, as this has many risks associated with it. Such risks are unfamiliar vendors, inappropriate software/hardware/service partnership, and discouraging potential vendors from committing adequate resources to their tenders for small winning chances, along with the high costs associated with the tender processing.

It is good policy to identify appropriate potential suppliers through the Request for Information, Expression of Interest and/or Request For Proposal, and notify them at the earliest opportunity that they may be asked to submit a tender. It is useful to notify the preselected suppliers of the tendering timetable, so that they can plan their own resources if they decide to submit a tender. The factors to be considered for preselection of a supplier might include:

- vendor's general background, such as history, business interest, location, and business status
- vendor's GIS background, such as experience, sales figures, client base, operational structure (marketing/sales, R&D, customer support, management), key vendor project/implementation personnel, and so on
- vendor system/service references, such as existence of user group, experience in similar configurations/applications/customer

Experience suggests that to invite four or six suppliers to tender is probably adequate. With this number, the chances of success for each supplier are sufficiently high to justify the considerable expense of tendering, while assuring a degree of competitiveness during tendering.

4.5.3 Preparation of the Request for Tender

A major risk in preparing the Request for Tender (RFT) is to be so specific in describing how the solution will be delivered, that suppliers are forced to offer suboptimal proposals simply to meet the stated requirements. Risk, on the other hand, is to have the RFT so flexible that the tender evaluation is near impossible. It is always worth remembering that the established suppliers have had many years of experience, and can often contribute innovative solutions if given sufficient freedom during tendering. To give the suppliers adequate freedom, it is appropriate to describe the basic GIS facilities that are required, and to describe the existing IT infrastructure into which the GIS must fit. For example, specify that you prefer to run a workstation network as a minium requirement of the system, but it is not suggested that you specify how the workstation should be configured.

The RFT should give as much information as possible to allow a comprehensive bid to be prepared, yet this should in no way attempt to hide requirements or conceal the extent of services to be provided. For the success of the project, both the supplier and the users must be in a position to win. Forcing the supplier to make an unrealistically low bid, with inadequate provision for development and support, is a high *long-term risk* for both parties.

Functionality, quality, cost, compatibility, reliability and the possibility to expand and upgrade the system should be considered carefully in the RFT. A standardized format for the tender will facilitate evaluation and ensure that the tender is complete in all essential aspects. The following is an example of the items that should be included:

Tendering conditions

This section defines:

- general tendering conditions (should) resolve any arguments that may arise and protect the organization by relevant law
- date and procedure for submission of the tender
- issues with prime contractors and subcontractors
- all important terms that may cause confusion (should be interpreted)
- all issues related to tax, fees, currency kind, exchange rates, and so on
- issues related to insurance, copyright, and so on (should be specified)
- issues related to failure to carry out the implementation on schedule
- issues related to the interaction with the organization during inspection, and installation (should be specified)
- other issues

System overview

This section will explain the history and the objective of the project. If it is appropriate, the current system, and existing equipment and new equipment required, can also be described.

System requirements

Detailed requirements, including software/hardware for data capture, data manipulation, data analysis and data presentation, should be defined.

Installation, acceptance, maintenance and other services

This section provides a preferred implementation timetable, and the resources available within the organization, internally, to facilitate the implementation process. It defines the condition of installation, testing and handing over, performance evaluation, provisional acceptance and final acceptance. It defines that the tender should demonstrate that equipment supplied is well maintainable on the operation site, and should provide options for maintenance agreement, post-installation support and training support.

Schedules of prices

This section requires the tender to supply the timetable for implementation, and the schedule of tender prices, maintenance charges, consulting and training charges, system upgrading charge and other relevant charges regarding this project.

At the same time, the evaluation criteria identified at the beginning of the preparation and outlining the selection process that will be used should be developed. The method of conducting the evaluation should also be determined at this point.

4.5.4 Issuing the Request for Tender

The RFT must be evaluated by the technical committee and approved by the steering committee before being issued to selected suppliers.

Preparing tenders is costly and time consuming. It is only reasonable to give suppliers sufficient time to complete their work to an adequate standard, and to have had an opportunity to resolve any issues with which they are unsure. Experience indicates that a period of around 6 weeks should be allowed for tender preparation.

Adequate resources should be set aside to answer any queries raised by the suppliers in a prompt and efficient way. It is unlikely that the RFT will describe the required system in such clarity that there are no areas of doubt in the suppliers mind. A *formal clarification meeting* is a useful forum to resolve these problems. To inhibit the clarification process is to risk tenders based on misconceptions and oversights. These errors may return to haunt the project at a later date. Arranging for at least a full half-day meeting between the project team and the suppliers' bid team within a fortnight of the issue of the RFT gives adequate time to resolve any issues that may arise before the tender deadline.

4.5.5 Tender evaluation

The mechanism for evaluating tenders should be considered at the time the RFT is prepared. The final fine-tuning of the evaluation criteria should be carried out, based on the information exchange with all suppliers. The tenders should be requested so that finding the relevant information for assessment is straightforward. Use of standard schedules for responses, boxes to be ticked, prepared itemized forms for descriptions of tendered item, and so on, all help to ease the task of evaluation.

There are several ways that tenders can be evaluated on an objective basis, and scores allocated to rank the various proposals. It is reasonable to generate objective scores in the following categories:

- initial costs
- long-term costs including future expansion and maintenance
- claimed compliance with requirements in the RFT
- corrected compliance with requirements of the RFT
- compliance with overall IT strategy

In addition, it is useful to generate scores based on a subjective assessment of the supplier. These might include:

- understanding of requirement and quality of tender
- technical advantage associated with proposals
- amount of customization needed to meet requirements, and so on

As soon as all tenders are received by the deadline, the first round of evaluation begins to eliminate the tenders that have not responded in essential areas. In asking suppliers to commit their resources to the tender preparation, they must expect a fair evaluation in return. To ensure fairness, the evaluation is best carried out by a team of at least three people working independently, but within preagreed guidelines. There should be no significant variance between the objective scores reached by each member of the assessment panel, though experience suggests that close examination of the costs may be needed to see what has been omitted from the tender.

A briefing note outlining how the subjective scores are to be calculated should be agreed by the team before assessment begins. It should be an essential part of the assessment process that the team must be prepared to justify their scoring to an independent review, and ultimately to the suppliers themselves, if required. Again, if scores are found to vary widely, the reasons must be established and a consensus found.

During the evaluation phase, each panel member will form an overall impression of the tender: its strengths, its weaknesses and areas where further information is needed in order to fully understand the proposal. Again, these should be documented as they will make an important contribution to the justification of the selection process. The RFT should ask for issues to be described that affect the long-term risks associated with the project. For example:

- the degree of user input requested
- the timetable for the development work
- the expansion capabilities
- the future development strategy of the supplier

- the reference sites with similar applications, and so on

The selection of two or three suppliers for further evaluation should be based largely on the ranking derived from the scores, but with overriding constraints, such as costs and supplier credibility, applied to modify the ranking, if necessary.

Once the final short list of two or three suppliers has been agreed, it is reasonable to inform all the other suppliers of their rejection. It is also a good policy to hold a debriefing meeting with any of the failed suppliers, at which the reasons for the rejection can be explained.

4.5.6 Resolving outstanding issues

With only two or three suppliers left in the tendering procedure it is time to resolve any outstanding concern about the proposals received. At the same time the remaining suppliers should be issued with the description of a benchmark test to be conducted in a few weeks time.

Inevitably, there will be a number of matters where the description given in the tender was inadequate to fully understand particular features of the proposed system. Questioning can be carried out through a combination of written exchanges and formal clarification meetings, where the supplier fields a technical team to address specific issues. During this phase, there are ample opportunities to assess the professionalism and technical competence of the staff of the suppliers who are involved. It is important to explore the outstanding risks and secure assurance on such issues as:

- the resources needed to finalize the system design
- the resources the supplier intends to invest
- the resources expected from the users
- a detailed project installation timetable
- timetable for the users becoming sufficiently experienced to contribute
- the true capacity of the proposed hardware and software

- the suppliers' development programs
- any problems with the proposed contract terms, and so on

During this phase, the project team and the supplier's staff will begin to form a working relationship, which will form the basis of the future implementation, should that supplier be successful.

4.5.7 Benchmark testing

The purpose of the *benchmark* test is to confirm that the claimed functionality and quality of the system both exists and can be demonstrated. The benchmark test should concentrate on those features of the GIS that it is essential to have in place, in order to support the major business activities identified during the business-analysis phase. In some cases it is unreasonable to expect the supplier to have created a system that meets all requirements of the RFT. It is essential that selected staff (users) attending the benchmark test are aware that they are to be shown the principles of each system, rather than the final product to be delivered.

To ensure that the benchmark is not reduced to a 'show business spectacular', it can be useful to use *your own data* as much as possible and *your own procedures*, and let *your own staff* be involved in the process.

It is important to record the views of the benchmark team at frequent intervals, to compare understanding and agree where further explanations and demonstrations are needed. It is surprising how the perception varies between team members of what has been shown during the tests. Additional tests should be requested if there are major concerns that have not been addressed by the other tests.

4.5.8 Final selection

After the benchmarking test, it is often found that there is rarely a clear winner in the final stage. Each system will have perceived strengths and weaknesses, which will complicate the final decision. It is useful to list the strengths and weaknesses of each system, and to discuss each weakness in turn to establish its effect on the overall project and the potential risk it introduces. The nature of the risks could be numerous, for example the effect on *operator acceptance* of an unfriendly user interface, the difficulties of *system and data management* and the long-term commitment of the supplier. There is no easy way to relate all these factors other than a thorough airing of the topic coupled with experience. In a few cases, the risks may be quantifiable in terms of lost revenue or upgrade costs, which could significantly impact on the long-term project budget.

It has been found appropriate to conduct a contract negotiation with all final players. This should be done with specialist legal input. It affords an opportunity to ensure that all supplier submissions and representations will be contractually binding. A *sell constructed contract* will minimize risks to the project by clearly identifying the roles and responsibilities of all parties. Particular matters to be mentioned in the contract could well include:

- guaranteed system performance
- guaranteed levels of development and support effort
- timetable for implementation, showing commitments from all parties
- levels of service for maintenance and support
- terms of payment and retention policy
- guarantees on migration to new or improved products
- ownership of intellectual property rights
- present and future discounting policy

The contractual details are the province of skilled contract negotiators, but heavy involvement from the key members of the selection team will ensure that the important issues are covered in the contract to reduce overall project risk. The completion of the contract can be a time-consuming process, but there are no short cuts. The discipline of producing a comprehensive formal contract focuses attention on the main areas of risk, and raises understanding of the commitments of both the user and the supplier to deliver an effective GIS on time and to budget.

Now is the time to select the tender by the technical committee. This decision must be supported by the results of the selection processes, and be presented to the steering committee for its approval, signature on a contract with the selected supplier, and notification to the remaining unsuccessful tenderers for the selection.

4.6 System development

After GIS software/hardware have been selected, the system development and pilot project begin. System development includes physical database design and operational procedures development, while the pilot project involves constructing the GIS database for a small representative portion of the project area, and refining the detailed database design. Internal/external database design professionals will play a very important role. Managers and users may at any time be asked for cooperation, comment and suggestion.

4.6.1 Physical database design and development of operational procedures

The physical database design is the highly detailed, logical definition of each individual data, and all of the data relationships. These are the specific and exact data and data relationships that must be converted or constructed by the conversion team.

Most specific geographic information systems support standard industry applications. Additionally, all geographic information systems have unique or special tools to enable enhanced operations to be undertaken in certain application areas. These standard applications and special tools need to be known prior to the design of the physical GIS database. GIS software normally provides full tools in spatial database design. But in terms of an attribute database, geographic information systems do not provide a database management tool as sophisticated as those of commercial database management systems. GIS databases today are almost universally designed using commercially available *relational database technology*. These commercial relational database management systems are the basis of the distributed computing environments that are so popular today. The use of a good commercial relational database is important, to be certain that investments in data will be levered over decades.

Conversion experts, especially those who have vendor-specific conversion experience, can make a big contribution to the physical database design (along with database designers, users, managers, and so on). The conversion expert should be a part of the physical database design team.

Initial automatic operational procedures should be carried out together with database design. It may focus on assisting data capture, such as customizing menu and interface of software, creating symbol library and automatic processing macros and batch files, and developing data interchange programs.

4.6.2 The pilot project

A *pilot project* is the limited-term use of a geographic information system, using data for a small geographic area to test the planned applications, to demonstrate the capabilities of the planned applications, and to demonstrate the capabilities of the system to key people in the organization.

The selected area should also be a miniature of the whole project area. The pilot project area normally selected should be around 5% of the total project area. The area must be the most difficult part of the project perceived by the database designer and should have a high diversity of data types, high density of data, elaborate processing and a full range of application.

Through the pilot project, the designed database could be refined to correct some mistakes, and improve the database structure so that the database is integrated, flexible and secure, with high performance and ease of operation. It is suggested that every opportunity should be taken to refine the database design before data capture with the entire project area, because any change in database design later will prove to be costly and time-consuming.

Typical pilot objectives (Montgomery & Schuch 1993) to be accomplished include the following:

- test database content
- test suitability of sources
- test database structure
- test document preparation and scrub activities
- test data conversion procedures
- confirm project-specific symbology
- test quality assurance procedures
- test data acceptance procedures
- test pilot applications
- confirm data conversion cost estimates/ budget
- provide a first major milestone for the GIS implementation process

CHAPTER 5

Managing the operation of a GIS

While the *acquisition* of a GIS is the transaction of technology, financial commitment and organization culture, the *management* of a GIS operation is meeting the challenge of the transaction and realizing the benefits brought by the transaction. Several important issues should be addressed in the stage of GIS operation, including data capture, data maintenance and system maintenance.

5.1 Data capture

5.1.1 Organization issues

Steering committee

The role of the steering committee is gradually downgraded. Initially, it may still have influence on management in guiding the transfer from implementation to operation. The responsibilities include, but are not limited to, the following:

- the balancing of the responsibility and interests among different parties

- financial management
- long-term planning

Management unit of GIS operation

The responsibility for GIS management is normally assigned to a department. For example, in local government in the United States the special GIS unit in the planning department, the taxation department, the cadastral department and the public works department are the major departments for GIS management. In the state of New South Wales the pattern is similar, with the engineering departments (public works) and the planning departments being the major sites for GIS management (AISIST 1993c).

Technical committee

At the data-capture stage, the technical committee functions as a coordinating organization to address short-term and operational issues, such as:

- strategy to build a GIS database
- training of staff to service and use the system
- standards in operation
- database maintenance
- information sharing
- cost sharing
- benefit sharing
- technology upgrade
- liaison with other organizations, and so on

Staff for data capture

Production manager

The production manager has primary project responsibility. This role is referred to as the *manager* or *operations manager*. The role is to accurately estimate data conversion and/or maintenance costs, margins, production or operation timing, and work schedules. Poor estimates often bring conversion and operation projects to a halt regardless of the competency of other staff members.

The production manager must have the ability to accurately *track* the production process and employee productivity. Tracking helps the manager to quickly identify project bottlenecks and hurdles so they may be overcome without reducing output for the project. Moreover, the production manager must effectively communicate and manage changes with staff members, clients and/or end users. It is the production manager's job to make sure that *the conversion and operation specification document* and *acceptance criteria* for each project are adequate to produce accurate GIS products, within budget, and on schedule.

Database integration/management expert

When a conversion or operation requires data integration with other enterprise databases or data export, a specialist is required. This person has specialist knowledge about how to use the data resident in other corporate files and to import the data to reduce overall data conversion costs, and how to export data to make the digital data directly available for other organizations.

Data conversion staff

Conversion staff often make up the majority of the GIS project team if the data capture is undertaken in-house. The role of conversion staff is to convert the GIS data on time, within budget, at the level of accuracy required. The conversion staff is also responsible for recommending the creation of *macros* or *programs* to speed up the process, and needs to understand the various methods of effectively dealing with source errors and record anomalies.

Scrub/records preparation staff

GIS projects often require the preparation of source materials or compilation of data from multiple source materials. This process is commonly referred to as the 'scrub' process. Such tasks can be performed directly on paper documents, or on-line in a raster or vector computer drawing file.

The scrub/records preparation expert is required to perform tasks such as feature layout and coding, as required by the project conversion specification. Source and document interpretation helps.

QA/QC and data acceptance staff

The specialist for quality assurance and quality control should define and manage GIS data *acceptance* plans and *quality control* procedures, and train the staff. This expert should be responsible for managing the final product to ensure it meets the conversion specification and data acceptance plan.

Other staff should be appointed, such as a computer system administrator, according to the progress of the project and the arrangement between the GIS unit and information technology (computer) unit.

It is required that all staff have appropriate training, if they do not already possess the required skills.

5.1.2 Strategy to build a GIS database

Strategy to build a GIS database (Castle 1993) involves total capture, phased capture, incremental capture and raster capture.

Total capture

A traditional approach in building a GIS database is that the data is converted and entered into the system, but is not usable until most or nearly all of the data is loaded into the GIS. This approach to conversion is a result of the notion that the best way to implement a GIS is to select the best hardware/software platform, identify applications and design a data model, and then perform the conversion to populate the entire database prior to GIS operation. By some, this approach is still thought to have the lowest cost and greatest accuracy, but it usually means a long period of time from GIS inception to the day of first benefits. There are alternatives, however, that offer acceptable functionality for the selective GIS user and they produce early data products.

Phased capture

With the phased-capture approach, one or a few parts of the conversion are completed and put into service. This enables a fairly quick gain of limited GIS benefits. It is intended that this is followed by additional phased conversion, funded by the benefits of the first phase. In theory, each phase of conversion is funded by the benefits of the previous phase.

Practical experience has not proven this to be a popular approach. However, for an organization with a commitment to a full GIS implementation over a longer time line, this approach may be beneficial to resolve a single, high-priority problem requiring an urgent solution.

Incremental capture

The incremental-capture approach is quick conversion (either in raster or vector format), by loading the GIS database with data covering a limited geographic portion of the user's total geographic area of interest. Full conversion, based on user demand, is scheduled or performed as a maintenance activity. This approach increases the cost of conversion, extends the schedule for full implementation, and diminishes the chances for high-quality data.

This approach produces maps for the computer very quickly, and spreads the high cost of full conversion over many years. In purely monetary terms, this approach negatively affects the cost–benefit ratio, as it increases both the total cost and the payback period. Immediate functionality may, however, override the long-term monetary considerations.

Raster capture

The raster-capture approach involves scanning the maps and data, attaching a reference geocode and creating a database. This is the quickest of all approaches, but produces useful results only in a few situations. It has not been widely used to date, and analysis and maintenance of raster data are difficult.

This approach may be used more in conjunction with commercially available databases where single-purpose solutions provide a high benefit. Some experts see this as a way to create 'one-shot' or 'throwaway' databases.

5.1.3 Project preparation

Clean-up source material/data

Before any data conversion begins, it is necessary to have a full check of the source material and data. The URAL report may indicate the availability of source material and data in general, but a detailed inventory should be undertaken (see Table 5.1).

Development of a data-capture specification

Data capture, whether it is undertaken internally or externally, requires the preparation of a written specification of requirements. It is this document that will set down all of the rules to be followed by the digitizing staff when capturing data, including data-capture routines, feature placement rules, symbol selection rules, quality assurance rules, and so on.

Data-capture contract

Whether capture is done *in-house* or purchased from a *capture service vendor*, there must be an agreement between the organizations involved

Table 5.1 *Data sources inventory*

	Spatial		Attribute	
	A*	D**	A	D
Coverage	Y	Y	Y	Y
Size	Y			
Quantity	Y	Y		
Material	Y			
Scale	Y	Y		
Projection	Y	Y		
Version	Y	Y	Y	Y
Accuracy	Y	Y	Y	Y
Custody	Y	Y	Y	Y
Structure		Y		Y
Format		Y		Y
Source	Y		Y	
Others				

* Analog
** Digital

that fully describes all the tasks to be performed. The interdependencies, participation and responsibilities of each must be clearly stated. Any agreement must include the means for qualifying and quantifying the work content. Compensation must be clearly defined, as well as realistic acceptance standards. Resources must be devoted to solve problems. If plans are not followed, this should also be documented.

At this stage of the proceedings, a corporate solicitor should be enlisted to help draft the contract. However, the following items should be included (Castle 1993).

Description of final products

A description of the physical format and the data specifications that should be delivered, is as follows:

- physical format
  - computer media
  - hard copy
  - plots or maps
  - database reports
  - quality control reports
- data specifications
  - data structure (target system): logical/physical file structures; working units,

naming conventions; symbology, feature definitions; mapping standards; and positional accuracy
- database specifications
- data structure
- entity/attribute definition
- data relationships
- valid attribute values/default values
- critical/noncritical attributes

5.1.4 In-house data-capture process
Project startup
Develop specifications for software customization

To develop specifications for software customization, you need to:

- create data entry screen
- set up and test data entry routines
- create graphic data capture commands/menu
- create graphic data symbology
- create graphic data capture routines

Establish the control points

Control points play an essential role not only in digitizing and producing an accurate map, but also in data integration. They establish a relationship between the map coordinates and the units that are in the digitized maps. The location of central points are normally available on the map document in latitude–longitude values.

Normally, a master control file is built for all layers of information that would need to be digitized. Using the file, all the spatial data sets can be transformed for a specific area into the same coordinate system, no matter what scale, projection, and so on.

Map scale issues

Digital or digitized datasets have no scale. One can display and output the dataset in any scale. However, the source documents have scales. Large-scale documents are more accurate to depict real-world features than small-scale documents. If a 1:10 000 source document is being used for digitizing, and the output is in scale 1:1000, it does not imply that the accuracy of the map has been improved.

Digitizing

Digitizing converts the spatial features on a map into digital format. As mentioned previously, there are two conventional methods to digitize: with a tabular digitizer and bureau scanner.

It is very important to define the working procedures before starting any digitizing work. If a large project is involved, a manual for digitizing will become necessary.

Scanning technology and *raster-to-vector conversion* algorithms provide new opportunities for *coverage automation* through scanning. Scanning has proven more efficient than manual digitizing, especially for documents with many lines. Source documents containing lines representing linear features and boundaries can be scanned to create *raster structure data sets* with very fine resolution. The raster data set can then be vectorized to create a *vector structure data set*. The accuracy of the output dataset depends on the map scale of source documents and the scanning resolution.

The maps must be carefully prepared. This involves cleaning any black spots, removing any unwanted points and/or texts, and generally improving the quality of any poor lines.

Scanned data cannot be converted into coordinates. It cannot create a control point file as required for digitizing. One solution is to draw a cross for each control point prior to scanning. Such control points are then scanned along with other lines. The conversion of the cross-hair into control points is done after scanning.

Research is currently being undertaken into *character recognition* and *symbol recognition* to realize fully automated data conversion. If the maps are of a uniform content, if there are many maps in a common series, and if adequate time and money exist to develop the rules and train the processor, pattern recognition may soon become a usable technology.

For maps containing contours, coastlines or area data, most vectorizing can run fully automated. For more complex data, such as urban plans and culture data, there are facilities to run the software interactively. Most map digitizing tasks involve a combination of simple and complex data, often on the same sheet. In such cases, users can benefit from the combination of full automation in certain areas, and semiautomatic operation in others. Using either option, the digital data is assured to be accurate and complete, and would require only minimal further editing.

Data Import

There are many GIS systems now on the market. These products differ substantially in a number of ways, including design philosophy, intended application areas and level of topology support. All these factors influence the design of a product's internal data model, and may require data used in one system to be translated before it can be used in another.

In some instances, data translation is relatively easy and uncomplicated. In other instances, however, it may be very difficult and require additional data entry, significant pre- and post-processing, and rigorous quality assurance. The level of intelligence built into the data, and the unique internal data models of the systems involved, determine the ease or difficulty with which data translation can be accomplished.

A variety of GIS data translation programs are commercially available. But they are not universal, two-way translators, and can communicate only with the system that has opened its data format (it is only recently that some very popular GIS vendors have done that). Many commercial GIS service companies provide data translation services. These firms generally use proprietary software and have a reasonable experience base.

Commercial GIS normally has the capability to access a vast spatial database; for example, the databases created for Northern America, such as TIGER and DLG. There is little communication between commercial software, but most geographic information systems communicate with each other through some standard file, such as DXF (Data Exchange Format), ASCII and IGES (Initial Graphics Exchange Specification). In most cases, what they get is the nontopological coordinates. The topological relationship could be created with GIS functions, if required. A few typical data formats are illustrated as follows (*GIS World* 1989).

Digital Elevation Model

The Digital Elevation Model (DEM) has been in use by the United States Geological Survey (USGS) since the mid-1980s. It supports grid-type data and a single attribute for each cell within the grid. The coordinate points of origin for the rows and columns, and spacings, are located in the header file; rows/columns are sequenced south to north, left to right. The format supports the transfer of both Universal Transverse Mercator (UTM) projection and latitude–longitude coordinates. The format has been used widely for the exchange of elevation data.

Geographic Base File/Dual Independent Map Encoding (GBF/DIME)

The Geographic Base File/Dual Independent Map Encoding (GBF/DIME) was developed by the Census Bureau in the US in the early 1970s to record and analyze census tract and block address information of the standard metropolitan statistical areas (SMSAs). All data is formatted as single points or lines with nodes at each end, and left/right information. The format is simple and widely used; however, it requires the user to exchange a great deal of redundant data. Although the GBF/DIME allows for multiple attributes for each feature, the field is limited to one record (300 characters) per feature.

Initial Graphics Exchange Specification

IGES is being used extensively for the exchange of computer-aided design (CAD) and computer-aided manufacturing (CAM). It was developed by joint industry/government representatives in 1979–80, and adopted as an ANSI standard in September 1981. It supports free field formatting representation/plotting of the information. The format is limited for exchanging digital cartographic data in that it can handle only one attribute per feature.

Map Overlay and Statistical System

The Map Overlay and Statistical System (MOSS) format was originally developed by the US Fish and Wildlife Service as part of the MOSS GIS. It is a nontopological format for vector data, and can be translated to, and from, a number of common spatial data formats. It is currently in use by several federal and local government agencies in the US.

Data Exchange Format

The DXF system is in a spatial data format, which was developed by Autodesk for its popular AutoCad system. The format is also used by several mapping programs that are based on AutoCad.

Tag Image File Format

The Tag Image File Format (TIFF) has been developed by Aldus Corporation, Microsoft and other commercial companies for raster data transfer.

Attributes capture and correction

Attribute capture uses the data entry technology of DBMS. Friendly interface, limited data size and permitted type for each field, plus default data setting, can improve the conversion efficiency, and help reduce the error of attribute data entry. The human/machine interface—the keyboard and the screen—is an area many believe to be ready for innovation. *Speech recognition* used in *voice data entry* seems to offer potential benefits in GIS conversion operations.

Voice input simplifies and enhances the input process by allowing the operator to concentrate on the data being inputed, rather than the input procedure. Theoretically, data is read in and commands are spoken. The system operates at the reading speed of the operator, which is much faster than the typing speed for most people. In addition to being faster, voice input has proven to be noticeably more accurate than keystroke input.

5.1.5 Quality issues
Accuracy and error in geographic database
Error is generally regarded as a measure of inaccuracy or deviation from the absolute truth. So while a measure of accuracy might define how correctly a feature is placed, a measure of error might define how incorrectly a feature is placed. Accuracy and error have four major parameters: spatial, temporal, descriptive and topological. The descriptive error is relatively easy to control, while the topological error is only an issue that may be raised when a topological data structure is used. Spatial error can be thought of as the inaccuracy of the true position of a feature with respect to the whole world, as determined by coordinates established through rigorous geodetic survey control points such as the corner points of real property. Sometimes the *relative accuracy* may be used to describe the variance that may occur when positioning two features with respect to each other. Figure 5.1 illustrates the typical spatial and temporal accuracy for some geographic data.

Source data error
Geographic data is inherently inaccurate due to the traditional methods of representing features with maps. This comes from:

- the original cognitive and scientific abstraction used by the surveyor
- the skills of the surveyor
- the precision of the instruments being used
- predefined policy on generalization, classification, symbolization and feature recording
- predefined survey questionnaires, description, and so on
- skills of the cartographer

- media for maps
- scale of map, and so on

But the accuracy of source documents (mostly maps) is normally controlled by standards issued by the national agency in many countries in order to guarantee the quality. The accuracy can be measured according to the standards. The most important factor that controls the accuracy of maps is the scale. Normally, smaller scale maps may have less accuracy. Table 5.2 is an example based on National Map Accuracy Standards (Montgomery 1993).

Table 5.2 *Map scale versus accuracy*

Map scale	Accuracy
1"=50' (1:600)	± 1.67' (0.5 m)
1"=100' (1:1200)	± 3.33' (1 m)
1"=200' (1:2400)	± 6.67' (2 m)
1"=400' (1:4800)	± 13.33' (4 m)
1"=2000' (1:24 000)	± 40' (12 m)

GIS processes and errors
The data-capture processes only exacerbate the errors in the source documents. The errors build up within the database through the various processes of data acquisition and manipulation. Some of the factors that determine the accuracy of the 'table digitizing' and the 'scanning' process are as follows:

- Source data
  - the accuracy of the source data
  - the scale of the source data
  - cartographic quality of the source map
- Staff capability
  - eyesight
  - hand movement and dexterity
  - carefulness
  - knowledge of the system
- Sampling techniques
  - spacing of the sampling frame
  - the spatial sampling rate
- Equipment resolutions
  - sensitivity of the detector
  - signal to noise ratio of the detector

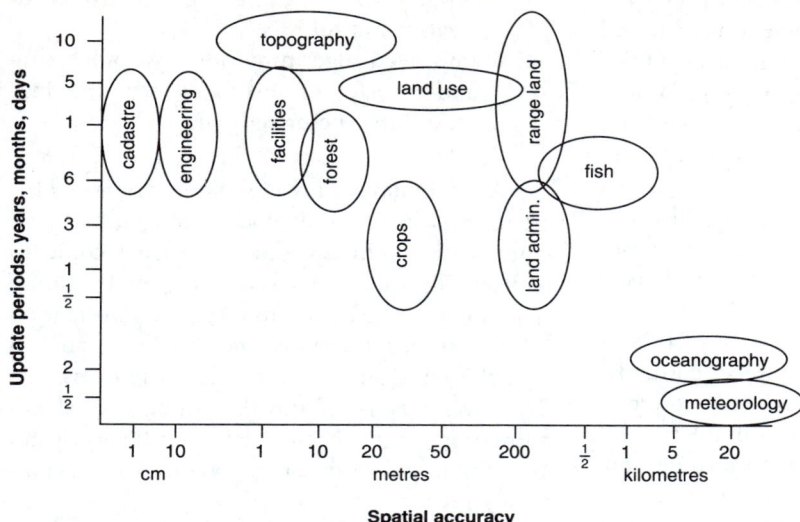

Fig 5.1 *Spatial and temporal accuracy*   ADAPTED FROM LYONS 1993

If topological data structure is used, there are routines with which one can build the correct topological relationship for a spatial database. For example, if polygon data is to be used, there is always a stipulation that the lines which define the boundary of the polygon meet exactly at the start and end point. Failure to do so at the data-capture stage requires editing. These usually rely on the use of a *user defined tolerance* to clear up problems of *nongeometric matching* in positional spatial data. If lines fail to meet at a junction, they will need to be extended, or the node points themselves will need to be moved, so that they join.

Error associated with data integration

The central role of a GIS as holder of spatial information is the ability to both link and integrate that information. Unfortunately, spatial, topological, temporal and descriptive error can all happen during this process.

The most common errors can be found through visually overlaying different map layers. If common features exist in two layers, and the data is from two source documents, then one may find spatial error and topological error. The most common example is the *sliver polygons*,

when lines that should be identical overlap each other, thereby producing small area mismatches; for example the easements may show on the wrong side of the road.

Error associated with data interpretation

Data interpretation from end users is also an integrated part of a GIS operation. Errors can occur in this process due to:

- lack of skill at data interpretation
- physical defection in eyesight, such as color-blindness
- misleading data and so on, such as improper data classification

Quality assurance

Total Quality Management (TQM) technology could be applied to quality assurance. While there are no fixed rules or simple steps for instituting TQM, a number of common elements have emerged that could be referred to as *principles*. The first of these principles is that anything that you do in your organization can be improved. Work consists of processes, and it is the processes that can always be improved. By their very nature, processes contain variation, and it is the variation that should be reduced.

The second important principle of TQM is that it is most often the processes that fail and not individuals. Dr Deming, founder of TQM, maintains that approximately 85% of all failure, waste, error, reworks, and so on could be directly attributed to process failure and not to individuals.

Another important principle of TQM is the concept that 'quality pays, it does not cost'. Improved quality will lead in turn to increased customer satisfaction.

The application of TQM to GIS operations implies that in order to do the project correctly, the process needs to follow every step correctly from data collection, data conversion, data integration, data maintenance to data interpretation.

At the stage of data collection, source data should be good quality and with accuracy higher than, or at least equal to, the quality required. Other specifications such as scale, projection, coordinates, version and producer should be well documented for later data integration.

During data capture, the error related to digitizing is largely unpredictable and random in nature. Every effort needs to be made to reduce the error to the minimum through effective training, rigid policy on how the data will be prepared and entered, and rigorous checking as part of a post-digitizing routine. Other measures include the introduction of improved incentives, and ergonomically sound environments for those involved in the digitizing process.

Some predictable errors that result from the equipment and the operator's physical defection could be detected through a benchmarking test for the equipment and aptitude tests for operators.

Scanned output cannot guarantee perfection, but solutions to the problems of accuracy and error associated with scanning methods of data capture essentially relate to three areas, as follows:

- the 'technical' improvements pertaining to scanning resolution, sampling rates and the speed of the process
- factors relating to the preparation and selection of material involving a human

input and requiring rigid patterns of quality control
- improved algorithms for post-processing and transfer of rasterized scanned material into vectorized data

As technology provides faster and more efficient ways to use and process geographic data, many data producers and users have come to realize the value of exchanging and sharing spatial data. The ability to integrate information from many resources depends on being able to find the data, and to understand the characteristics and quality of the data. The data documentation for both data indexing and assessment purposes, or data about data, is commonly referred to as *metadata*.

Normally the data elements should include the following:

- *Identification section* describes the general data content, spatial extent and use of the data.
- *Projection information* describes the horizontal and vertical coordinate systems used for the spatial data.
- *Data custodian information* provides points of contact for the data.
- *Access information* provides details about the means of, and conditions for, accessing the data.
- *Status information* describes the state of maintenance cycles for, and policies on, availability of the data.
- *Table definition portion of the data dictionary/schema* describes the thematic content of features portrayed in the data.
- *Table attributes portion of the data dictionary schema* describes the thematic content of the characteristics of the features portrayed in the data.
- *Source information* describes documents used to compile the data.
- *Processing steps* describe the procedures and parameters used to convert the source materials to the final data.
- *Data quality* describes the quality of the data to assist potential users to determine

whether the data is suitable for an application.

- *Metadata reference section* describes the currentness of the contacts for the metadata.

Solutions to errors associated with *data manipulation* could be as follows:

- When a variety of data sources are to be used, rigorous checks as to the quality and applicability of the data should be performed.
- Alternatives to utilizing a tolerance based upon geometric values rely on utilizing features and attributes, as a means for sliver polygon identification and removal.
- Alternatives to reassignment of space by geometric attributes can be found; however, they rely upon utilizing a user's knowledge of the data, and as yet they are not offered as a standard feature of any large GIS.

5.2 GIS data maintenance

5.2.1 GIS database maintenance

GIS database update

Any user of a GIS that involves updating data over a significant period of time, whether updating spatial data or attribute, is engaged in a long transaction (Newell 1994). Contrast this with the user of a commercial DBMS application, such as banking or airline reservations. In such an application, a user may prepare an input screen over a period of a few seconds, which then updates the system, resulting in a transaction that lasts a small fraction of time.

The successful maintenance of a GIS database is dependent upon how effectively the GIS can handle the long transaction, since the update can hold up the system for a period of time even though it may be running as a stand-alone system or in a network. Commercial DBMSs can handle transactions that are open for a few seconds or minutes, such as attribute data update, but certainly not those that last for hours or days of spatial data updating.

Newell (1994) illustrates one solution: the user who wishes to update the database requests that for that system the part of the database on which he/she wishes to work be copied into a single-user database. Whether or not the single-user database is proprietary or is a commercial RDBMS does not matter, as it handles temporary data only. If a database can be held on the local disk of a workstation, then the user places no load at all on the database server or the network while he/she is working. However, the disadvantage is that it is difficult to maintain relationships between the data that is downloaded and the data that is not.

Newell proposes another approach, that is to implement a mechanism for *version management* deep in the database engine itself. A version managed database is capable of holding any number of versions of the whole database without replicating data that is common between versions. Thus, all users can see the whole database at all times, subject to any changes being made within the privacy of their own versions.

Good management practice can remedy the shortcomings of the technology. For example, update practices can be coordinated:

- among different maintenance staff
- within the department or among different departments
- between the attribute database update and attribute update
- among the different attribute database updates

Coordination of the activities between the GIS user and maintenance staff make sure the users are fully acknowledged by the updating.

Responsibility for database maintenance

GIS hardware and software are normally looked after by a special GIS unit or by the general IT department of the organization. But the maintenance of GIS database is quite complicated. Since the GIS database consists of integrating various data from different departments, the issue is not as simple as the hardware and software.

Maintenance of a shared database would ideally occur in a uniform, standardized and timely fashion, regardless of whether it was centralized or departmentalized. Successful GIS maintenance can occur in a departmental fashion, with two or more departments providing updates to a *shared database*. Some systems are initially designed for *source updating*. This type of maintenance relies upon set standards and maintenance applications, along with uniform procedures, to perform routine maintenance to a shared base map.

Another method is the formation of a separate maintenance group in the organization, which is in charge of all spatial data updating.

Cost for system maintenance

Provision needs to be made in the overall running cost budget for the maintenance cost of the system. This can be made by one or a combination of the following:

- shared department budgets
- transaction proportion in the budget
- overhead allocation

Data maintenance schedule

Detailed GIS data maintenance procedures are required, no matter who is responsible for the data maintenance. In preparing the procedures, personnel training, quality assurance and other related tasks should be considered. Forms may be developed to track information such as operator's name, type of updates made, job or project numbers, and so on. (Montgomery 1993).

5.2.2 Marketing GIS data

The demand for data from a GIS database by an external organization must be expected with most GIS projects, the marketing of the GIS data being through the sale of hardcopy of data and digital data.

There is often a problem in cost definition in digital data. The cost of duplicating digital information products, data or software is negligible compared with that of collecting and processing. The problem of selling the data is complicated in two aspects; the price of the data and the data ownership.

Price policy

Since most GIS projects are largely funded by different levels of public organizations, the price policies are even more complicated, because in most countries relevant legislation will prevent the recovery of the cost funded by public organizations (Soursa 1993).

The Policies for New Brunswick GIS are as follows:

- All types of customers will be treated identically.
- Multisite licenses will be used to serve major customers.
- Market-based pricing will include a volume discount policy.
- Provision will be made for a special class of customers such as developers.
- All products will be copyrighted.
- Licensing arrangements will be possible.
- The corporation may limit its liability.
- Fees will be set according to the product or service category.

The classification of products and services has been made into four categories:

- statutory products and services—those services that, by law, the organization is required to provide, subject to a standard fees structure
- existing infrastructure—those infrastructure products for which the production costs have already been funded, and which now require funding for maintenance and distribution
- future infrastructure—those infrastructure products for which there is an overall need, but which have not yet been developed or funded
- retail products—value-added products and services created primarily for general use, such as atlases and customized maps

The pricing approaches include:

- legislated prices, which will be set by regulation, and will be adjusted from time to time to accommodate inflation, level of service and other factors
- incremental cost recovery, which is the cost over and above the costs of developing the database, and which includes:
  - the cost of legislated prices
  - actual duplication, media and computer charges
  - any direct costs associated with distribution
  - a charge to cover maintenance of the product
- partial cost recovery, which is the assignment of a portion of the full costs of developing a system plus incremental costs, and which will be guided by:
  - the cost of incremental cost
  - a hardware and software allocation cost
- Market value pricing, which is the price set by market research that reflects the supply and demand characteristics of the produce or service in question:
  - the cost of partial cost
  - an opportunity cost of the net assets employed as set from time to time by the corporation

Pricing policy implementation

Beside the policy, the characteristics of products and fee schedule based on this policy are as important tasks as developing the policy itself. The demand of products, and the cost structure from site to site and project to project, vary considerably, and at present no guidelines can be identified.

5.3 GIS technology maintenance

The GIS *technology maintenance* includes the software development, software upgrade, hardware upgrade and the application of new technology, as well as cleaning and repair.

Hardware has the shortest life among all components of a GIS. The performance/cost ratio improves significantly over time. New technologies may emerge as the effective tools to improve productivity.

A GIS firstly is used for data processing. It is not a turnkey technology, an expert system or a decision support system. Migrating the GIS into a decision support system is an obvious direction for future development. It can be further adapted for the integration of GIS and operational, management and planning procedures.

The development of GIS software never stops. This is reflected by the increasing output of version numbers and independent modules that are continually becoming available. Normally, the latest version of GIS software will:

- create some bugs in an early version
- add on new functions
- improve the functionality performance of the previous version
- migrate to a new operational environment, such as operation system, hardware and new network
- improve connectivity with commercial DBMS, or other software package, and peripherals
- make the software more user-friendly, and so on

CHAPTER 6

Case studies

Introduction to case studies

Throughout the 1970s investment by local governments in Asian and the Pacific countries amounted to hundreds of millions of dollars. The local government association in the Netherlands had already decided from its experiences in attempting to introduce digital geographic information systems during the 1970s, that the whole process was not cost-effective with the hardware and software available at that time. Many other countries at that time found themselves in a similar situation.

The spatial information industry responded during the 1980s with the production of computers with a much greater data-handling capacity for a greatly reduced cost. Very considerable advances were made in the capability of the software and the peripherals.

Planners and decision makers have high expectations of the hardware/software systems that are now commercially available. However, local government and state/national government agencies are only now, in the mid-1990s, coming to the realization that this requires the application of an entirely new technology, which is still in the making.

Some agencies and enterprises have successfully installed parts of a total GIS that:

● meet their specific and recognizable needs

● have clear indications of the training of staff that was necessary

In the past, agencies, departments and enterprises invested heavily in the hardware, only to find that much more budgetary allocation was necessary for identifying the software and peripherals, and arranging for the specific training of staff to make the GIS, or, as it has become known in the 1990s, the Spatial Information System (SIS), cost-effective for their needs. The industry has now responded to the need with reasonably priced hardware, software and peripherals, and also the government agencies and private enterprises have spent time in specifying more precisely what they need the GIS/SIS to perform for them (that is, having an understanding as to what the SIS hardware/software can produce for them under very specific circumstances). Consequently, the GIS/SIS industry continues to expand with the production of hardware and software aimed specifically at their needs. Portraying the successful results of several major GIS/SIS projects within the Asian/Pacific region discussed in this chapter is fraught with limitations. However, during the first part of the 1990s, many examples of software packages for identifying the utilities layout of the towns and cities, land use notations and municipal engineering, housing inventory projects, forestry inventory programs and several other cases have become available.

Progress is being made towards the provision of software that will embrace the several interactive dynamic functions, which together go to make a city function successfully as a whole. But such software programs have only recently become available. Until a city authority or national government agency has installed the system, and has proven its performance with trained and skilled staff, it is somewhat early to write up successful case studies.

As a means of illustrating the present state of the art, three case study examples are included:

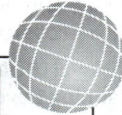

- Semarang, Indonesia: GIS for land resources management and regional planning
- People's Republic of China: Pilot Project for Zhuhai City Government, Guandong Province
- Australian Bureau of Statistics, Australia: GIS and Spatial Information System for mapping a human resources profile of the country

Case Study I

Semarang, Indonesia*

GIS for land resources management and regional planning

By N. Hariadi and J. Cameron, Curtin University of Technology, Perth

Introduction

Land resource management and regional planning are important and critical issues for the government of Indonesia. This is because economic development is based on the management and utilization of natural resources. On the one hand, there is a need to exploit land resources for improving the standard of living but, on the other hand, there is a need to conserve these resources for future use. GIS technology has been accepted as an important factor contributing to land resource management and regional planning.

Various efforts have been attempted to develop a GIS at the national and regional levels. The major project concerned with the development of a GIS at the national level is the Land Resources Evaluation and Planning (LREP) project. This project was started in 1985 and completed in 1990. Its purpose was to develop a land resources database and a GIS at the provincial level, in particular in the provincial development planning agency. The project was established in eight provinces of Sumatra and each provincial planning agency was equipped with a PC, plotter and ERDAS GIS software. Results from this project have been unsatisfactory. The implementation of GIS technology for regional activities in the provinces is still very limited because of the lack of regional information and the limited number of fully trained technical staff. Careful assessment of this project is now necessary and ways need to be determined to make it more effective, particularly as it was expanded in 1991 to other provinces.

---

* From Australasian Urban and Regional Information Systems Association (AURISA), 1993 Proceedings, 22–26 November 1993, Adelaide, pp. 128–36.

The Urban Land Information System (ULIS) project was another pilot project concerned with the use of GIS technology at the district level. This project was established in Semarang City in 1986 and completed in 1990. It was equipped with two PCs, a CalComp 9100 digitizer, a CalComp 1023 plotter and MicroStation Intergraph software. At present the ULIS does not work properly because of insufficient technology transfer between the consultant and local government staff. The ULIS is also very tedious and expensive to operate for the local government capabilities, (Soegijoko 1989).

In addition to these, there are many other GIS projects in Indonesia, although development has been slow due to many problems and issues that need resolving. However, because it is recognized as an effective tool for land resource management and regional planning, better methods of implementation, particularly at the local government level, must be used.

Development planning in Indonesia

National planning

In theory, development planning in Indonesia is a bottom-up approach, beginning at the district level and leading to the provincial level through the following units:

- district or municipal level (*Kotamadya/Kabupaten* or Level 2 in the provincial planning structure)
- provincial level (planning Level 1)
- provincial coordinating level (with up to five provinces involved)
- national level

At each level, 5-year development plans as well as annual plans are prepared. Regional development plans at district level should be capable of incorporation in the provincial plans, and the national plan should be based on the results of provincial plans. However, the regional plans at district level have little or no impact on the provincial plans, and provincial plans have limited impact on the national plan. The effectiveness of bottom-up planning has not been satisfactory.

One major constraint is that regional information is acquired in different formats, especially at the district level. The use of GIS technology is also very limited, because it is not well understood at this level. Implementation of the technology, using a standardized format for data input and output at district level is urgently needed to support bottom-up planning from provincial to national levels.

Regional responsibilities and problems

BAPPEDA is the regional development planning agency within local government. There are two levels of *BAPPEDA*: provincial and district levels (*Datti I* and *II*). The development of a GIS is designated only for *BAPPEDA* at district or municipal level (*Kotamadya/Kabupaten*). *BAPPEDA* is responsible for the determination of policy on regional development planning as well as its evaluation. It plays an important role in the management and utilization of land resources in a region. Unfortunately, most of the local planning agencies at the district level lack the necessary facilities and staff trained in land resources management and regional planning. Semarang's planning agency is no exception.

Land resource and other regional data are often out of date and contain serious errors because of staff and equipment limitations at the district level. *BAPPEDA* can only work with data, supplied or collected, that requires little change. This situation has worsened in the last 5 years because of rapid population growth.

Bad and insufficient data introduces many serious problems into the regional development process. Delayed development, goals not achieved, overlapping responsibilities and conflicting land use policies are obvious shortcomings faced by local authorities. There is an urgent need for effective methods to carry out land resources management and regional planning at the local government level.

Objectives

In broad terms the objectives of development programs are to increase the standard of living of the population and foster economic development. Nevertheless, development programs must be based on sound management and utilization of natural resources. Natural resources need to be exploited in a rational manner to meet the needs of future generations. This project examines ways to support sustainable development in a region using a structured approach in applying GIS technology for land resources management and regional planning at district level. This can be achieved in the following manner:

- building an integrated land resources database
- creating appropriate applications in regional planning and evaluation
- developing a flexible, easy-to-use *graphical user interface* for map production and database query

The study area

Semarang, the capital city of Central Java province, was selected as the study area because of the complexity of urban growth and conflicting land uses. In addition, Semarang was the pilot project area for the ULIS project, which was not properly implemented because of the lack of fully trained technical staff and poor coordination between local agencies. Semarang is also one of the priority areas for the second LREP project in Central Java.

Semarang is a small and crowded city with an approximate area of 374 square kilometres (144 square miles). It is divided into nine administrative subdistricts (*kecamatan*) and consists of 177 communities (*kelurahan*). In 1990 Semarang's population was about 1 146 931, with a population growth of 2% per year and an average population density of 3072 people per square kilometre (0.386 square mile). A number of urban areas, representing 30% of the area, include Central Semarang, West Semarang and the northern area of South and East Semarang. The rapid growth of new urban areas is due to population pressure. A large amount of agricultural land is being replaced by new settlement areas, and demands for new services and facilities cannot be accommodated as needed.

Building an Integrated Land Resources Database

An Integrated Land Resources Database (ILRDB) is a unified, multifunctional database. It is an integration of data requirements from all the users of land resources and regional data within the development planning agency. The development of an ILRDB considers all aspects of regional planning activities. These aspects, or groups of activities, include economic, sociocultural and physical aspects and infrastructure. In addition, there are two other groups that are functionally related to planning activities, namely research and statistics and reporting groups.

Following the functional requirements study, it became apparent that there were numerous requests for the same land resources and regional data. There was also quite specific data that was to be independently collected and used by each group of activity. The ILRDB is, therefore, the best fit derived from common land resources and regional data needs. Expectations are that the database will become an integral part of the operational, managerial and administrative functions of the planning agency, with each group satisfied.

A conceptual model of the ILRDB was developed based on two criteria:

- the integration between various kinds of land resources and regional information needs by each group, for the development of integrated land resources management and regional planning
- the elevation of existing data sources (maps) and other related information

Figure 6.1 illustrates the conceptual model of the ILRDB for the regional development planning agency at Semarang City. The ILRDB consists of six databases: base map, land, environmental, utilities and infrastructure, statistical and cartographic. Elements for each database, except for cartographic, are listed in Table 6.1.

The cartographic database contains all the data necessary to produce a map and linkage data, connecting the cartographic features with the corresponding map coverages or objects in the other databases (Frank 1991). These linkages include associated map attributes for display purposes, a look-up table, map label and symbols sets, and the macro-files or procedures to display maps, either on a screen or on paper. The cartographic database also contains a collection of map models that were built to provide tools for the production and reproduction of maps.

Application developments

The first task of the project was to place all the required data into a unified digital database. The sources of land resource and regional information range from topographic maps to thematic maps through to statistical data. PC Arc/Info Version 3.4D Plus was used as integrated GIS software to automate, manipulate, analyze and display the land resources and regional data. DBASE IV was also used as a relational database management system to manage and analyze the relevant thematic and statistical data.

The purpose of application development is to produce an integrated evaluation product. Land resource and other regional information can be combined and then analyzed to produce specific products that are *highly derived* information. These products will provide the planner and decision maker with information not only about basic classification or physical evaluated products, but also about implications for resource allocation decision. Integrated evaluation products are absolutely essential to make rational policies for regional planning (Fessenden 1984).

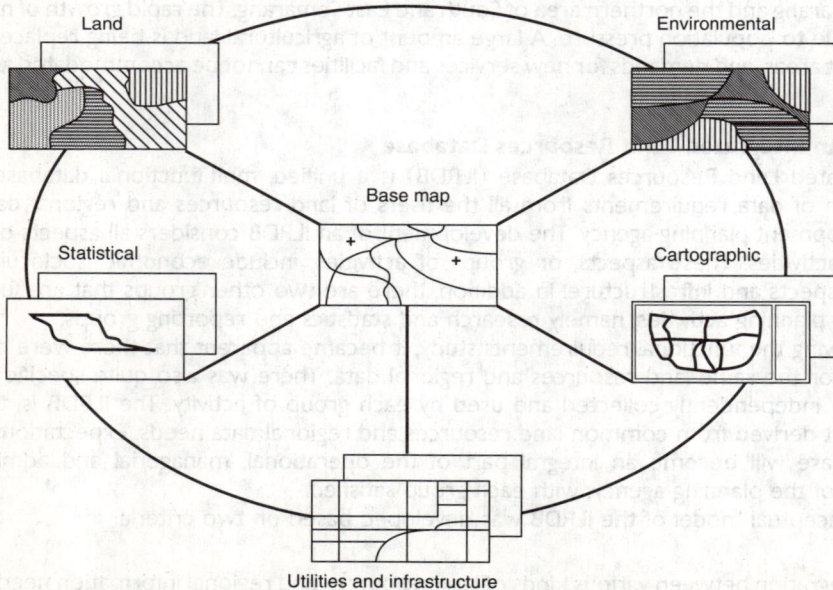

Fig. 6.1 *Semarang's Integrated Land Resources Database (from ESRI, 1990)*

Table 6.1 *Land resources and regional information needs of the Regional Development Planning Agency of Semarang City*

Database name	Thematic layer	Primary features
Base map	Hydrology	Streams and coastlines
	Roads	Road network
	Administration	Administrative boundaries
	Elevations	Contour lines and spot heights
Land	Land use	Present land use
	Land system	Land system, land suitability and environmental hazard
	City planning	City master plans
Utilities and infrastructure	Utility	Transmission, pipe and telephone lines
	Economy and industry	Distribution of market, industry and banks
	Infrastructure	Railways, airports and harbours
	Social infrastructure	Distribution of schools, hospitals and parks
Environmental	Slopes	Slopes and slope stability
	Geology	Environmental geology
	Geomorphology	Landforms and climatic range
	Ground water	Ground water surface
Statistical	Statistical data	Population, education, social, economic and industrial features

The establishment of land use policies for new urban areas is an example. The combination of present land use, land system, land suitability and environmental hazard provides the physical land evaluation for a planner to specify land areas that are suitable for urban areas. However, physical evaluation alone is not sufficient to make decisions about establishing these areas. Socioeconomic analysis must be incorporated into the physical land evaluation in order to develop rational policies for land use.

Comprehensive analysis and modeling capabilities are required to carry out large and diverse types of applications in land resource management and regional planning. It is expected that a wide range of GIS applications will be performed using the available database. The potential applications of a GIS in land resources management and regional planning at the regional development planning agency include area mapping and reporting, demography and human resources, development tracking, land use planning and environmental management, and locational analysis.

User interface development

The development of a user interface is another important element of the project. In order to provide more effective support for GIS use in regional development planning activities, two kinds of user interfaces were developed:

- a dynamic, user-friendly interface for map composition and database query
- an application-oriented user interface in land resources management and regional planning activities

A user interface was developed using Simple Macro Language (SML) in PC Arc/Info. SML is a procedural language interpreter, designed specifically for the customization of repetitive tasks and the development of GIS applications.

The provision of dynamic user interfaces for map composition and database query allows planners and policy makers to query the land resources database and produce thematic maps, either on screen or on paper, without the need to understand the complexities of a command language. Database query interface enables the users to display and interactively query any particular database and its associated attributes within the land resources database. Map composition interface is covered not only for the final map production, but also for the check plot. Thematic maps include present land use, land suitability, population density, administration boundary and many others. A composition interface for topographic or base maps to meet specific user requirements is also available.

An application-oriented user interface was developed that provides user-friendly tools to promote understanding and awareness of GIS benefits in land resources management and regional planning. This customized user interface provides the macros so that land use planning can be implemented. This interface includes land use assessment and planning, and locational analysis functions. In addition, it can be used as a guide for developing other GIS applications in land management and planning areas.

Conclusions

As a result of population pressure on agricultural areas, many problems have developed in resource management and planning at regional and national levels. GIS technology is seen as a timely and appropriate mechanism for resolving such problems, particularly in Semarang City. However, at present there are limitations in using conventional GIS methods because existing data is too difficult to access and staff are insufficiently trained in the use of the new technology. Therefore, existing GIS approaches need to be modified.

The research undertaken at Curtin University of Technology has identified a three-step process, which can provide effective land planning and land management results at the local government level, using GIS technology. This process involves the creation of a unified database using existing data sources, the identification of appropriate applications and the development of customized modules to implement these applications. The application of this methodology in Semarang City will need to be continually assessed before optimum results can be achieved.

References

ESRI 1990, *Understanding GIS–The ARC/INFO Method*, Environmental Systems Research Institute, Redlands, California.

Fessenden, R.J. 1984, *Natural Resources Data Base Project, Final Report*, Alberta Energy and Natural Resources, Resource Evaluation and Planning Division, Edmonton, Alberta, November, 37 pp.

Frank, A.U. 1991, 'Design of Cartographic Databases', in *Advances in Cartography*, ed. J.C. Muller, Elsevier Applied Science, London, UK, pp.15–44.

Soegijoko, B.T.S. 1989, Socio-Economic and Spatial Planning Bureau, National Development Planning Agency, Republic of Indonesia, Paper presented to the International Conference on Geographic Information Systems Applications for Urban and Regional Planning, Ciloto, Indonesia, October, 42 pp.

Case Study 2

People's Republic of China

Pilot Project for Zhuhai City Government, Guangdong Province*

By Lihua Li, Project Manager (Land Information Centre), University of New South Wales, Sydney

Project background

The Zhuhai Pilot Project has been developed as a consequence of the extensive scientific and technological cooperation between the Lands Department of Guangdong Province (LDGP), People's Republic of China, and the Department of Conservation and Land Management, Land Information Centre (LIC) in New South Wales, Australia.

China is a diverse and complex nation, rich in natural and human resources. The management of these resources has required the development of comprehensive management structures including the use of sophisticated management tools, such as Land and Geographic Information Systems.

The LIC and the LDGP have recognized that the implementation of this technology is not a solution in itself and, with the cooperation of the Zhuhai City government, have designed a project to analyze the requirements, and plan the implementation of a Land and Geographic Information System, within the Gongbei District.

The main objectives of the Pilot Project are:

1. to carry out the user requirement study of the organizations involved in the project
2. to produce a physical data model for *information system development* in the project area
3. to develop a pilot implementation of such a system
4. to develop a strategic plan for system coordination, management and implementation for the project area
5. to draw the recommendations and guidelines for Spatial Information System development for the Chinese municipal governments generally

For implementing the pilot project, the LIC provides hardware, software, CASE tools, technical expertise and project management, and the LDGP and Zhuhai City government provide technical support, infrastructure support, data-capture staff and project management.

The pilot area

Zhuhai City is located at the mouth of the Pearl River in Guangdong Province, adjacent to Macau, and approximately 40 kilometres (25 miles) from Hong Kong. It is one of the four Special Economic Zones (SEZs) in China proclaimed in November 1980.

---

* This case study was written especially for this book.

Gongbei District is one of four districts of Zhuhai City, and it is the area of an active land market with a variety of land users, from private individuals to city, provincial and national level government organizations. The total area of Gongbei District is 23 square kilometres (8.8 square miles), with a total permanent population of 46 000 (temporary population as a holiday resort is 125 000) as the data for the user requirement study.

The pilot area is within the Gongbei District and covers approximately 9.5 square kilometres (3.7 square miles). Zhuhai City is the only area for which the cadastral survey has been completed.

The Participants
The Land Information Centre has been serving the Spatial Information Science needs of New South Wales for over 40 years and is acknowledged as a world leader in the development and applications of Spatial Information Technology. The LIC's 400 officers present a unique, unparalleled range of skills, and are able to provide assistance on a number of levels, from the supply of information to the 'development implementation' of the most complex Spatial Information Systems.

The Lands Department of Guangdong Province is a government body directly under the supervision of the provincial government. Its main tasks and responsibilities cover surveying and mapping, land administration, land use planning, zoning, development, land legislation, supervision of policy development and implementation in Guangdong Province. It is also responsible for coordinating activities related to land renovation between different departments and regions.

The cooperative relationship between LIC and LDGP began in the sister-states relationship of New South Wales and Guangdong Province. A Memorandum of Understanding has been signed between these two organizations for science and technology cooperation.

The Zhuhai Lands Bureau, directly administrated by Zhuhai municipal government, and indirectly administrated by the LDGP, is responsible for the land management and planning policies, and the implementation of the policies and management activities, directed by the municipal government.

Project cycle and progress to date
The Zhuhai Pilot Project consists of the following components/stages:

- project preparation and planning
- user requirement study
- data modeling
- system development
- training the Chinese staff
- system commissioning
- data capture
- evaluation and recommendations

The project preparation and planning began in August 1993. The user requirement study commenced in November 1993, and was completed in May 1994 (due to Chinese New Year break). Data modeling started in June 1994 and is due to be completed early in 1995. The following stages are system development, project staff training in Australia, data capture and system commissioning in China. It is planned to complete the system development by February 1995, and to commission the system in March 1995.

Project preparation and planning
The most difficult problem confronting project managers is implementing the programs, and much of this can be traced to poor project preparation and planning. Project preparation and planning is the most crucial phase of the project. It defines the project scope, project management

structure, project methodology, quality assurance procedures and project implementation schedules to ensure efficient, economic use of resources, and to increase the chances of implementation on schedule.

Project scope/deliverables
Defining the scope of the project is important, but not easy, especially for the cooperation in the project when cooperating partners have different view points, and slight deviation from the objectives. Negotiation or discussion among the project parties is required, and agreements have to be reached, before going further, to plan the project.

It has been agreed by the project proponents that the Zhuhai Pilot Project will deliver the following elements:

- project plan
- data model of the existing system
- data model of the proposed system
- pilot implementation of prioritized components
- data-capture program and specification on tools provided
- strategic implementation plan
- training program

Project plan
The project team spent 3 months developing and reviewing the project plan. The plan covers the following aspects:

- project management structure
- project management methodology
- system development tools
- training strategy
- quality assurance procedures
- user requirements study
- data modeling
- pilot system development
- pilot area data-capture specification
- pilot system commissioning and evaluation

Project management structure
The project structure has been developed to ensure the project's success. The project is overseen by the Steering Committee, which is composed of the principals of the three participating organizations. It is responsible for executing the agreements, and ensuring that the relevant project activities are carried out by their organizations. Directly responsible to the steering committee are three project managers from three cooperating parties. They are responsible for running day-to-day activities that constitute the project plan, and reporting to the steering committee, as required. In addition, there are two quality assurance managers from Australian and Chinese sides responsible for the project quality assurance program. Each organization has also nominated three key technical personnel to carry out the activities required.

Project management methodology
Apart from a funding constraint, the Zhuhai Pilot Project has unique constraints; these are different culture backgrounds and communication barriers (long distance and language problems), and

each team member has to serve his/her core business functions first, which limits the availability of key staff.

In order to overcome the cultural and language constraints, the LIC has appointed a Chinese-born technical officer as its project manager. This project manager and the LIC's project team worked closely to develop a special project management methodology (a mixture of both Chinese and Western philosophy), called LIC Project Management Methodology (LICPMM), to best fit the project environment. The LICPMM has:

1. defined the project management structure, responsibilities of each level and participating organizations, and reporting and dispute resolution procedures
2. broken down the tasks into a set of subtasks, deliverables and deadlines, and has assigned the individuals to a single subtask
3. developed rigid quality assurance procedures for each stage and project review procedures

The LICPMM has proven to be an effective methodology during the project implementation.

Quality assurance
The quality assurance comprises the standards and procedures for conducting the Zhuhai Pilot Project, including the following elements:

- responsibilities
- project planning and requirement for control procedures
- design control procedures
- document control procedures
- configuration management
- process control procedures
- inspection and testing procedures
- project review procedures
- documentation and records standards
- dispute resolution procedures

User requirements study
A user requirements study commenced in November 1993 and lasted for 6 months in Zhuhai City. During this task, the project team conducted a *user workshop*, a series of key user interviews and a questionnaire survey.

User workshop
A 2-day user workshop was organized prior to the project on-site commencement. It is an important task of the user requirements study. The purposes of the workshop were:

- to introduce GIS application and the objectives and methodology of the pilot project to the key organizations to be involved in the project
- to raise the awareness and understanding of the project
- to gain support from those organizations

The workshop material was prepared by the LIC's project team in conjunction with AISIST. The workshop was conducted on 18 and 19 November 1993 by the LIC's project staff, with assistance from the staff of the LDGP and Zhuhai Land Bureau. There were fifty-three executives from key organizations attending the workshop. During the workshop, participants gained good understanding of GIS application and what Zhuhai Pilot Project will bring to them; they expressed their support for the project.

Key user interviews

Eight key organizations, including land, planning, transportation, utilities and city administration, were selected for the interviews by the LIC's project staff, with assistance from the LDGP and Zhuhai project staff. These interviews were driven by a set of briefing notes developed during the project preparation period, and will be designed to develop an organization model for:

1. organizational responsibilities
2. business requirements of the organization
3. high-level data requirements of the organization
4. overview of functional requirements

During the interviews, the project team provided, for each interviewee organization, a brief overview of the project and its intended goals, in addition to introducing the questionnaire and discussing the requirements for each organization to properly complete the questionnaire. Comments, and any discussion arising from this questionnaire, were used to modify the questionnaire prior to questionnaire distribution.

Questionnaire survey

The purposes of the questionnaire survey for the project were:

- to acquire data/information on organizational responsibilities, interrelationships with other organizations, business functions, processes, work flows, data sets requirement and details of attribute data
- to produce an ordered list of user requirements relating to the requirements for a LIS/GIS
- to use the information collected from the survey for data modeling and system development, to meet the users' day-to-day requirements

The draft questionnaire was designed by the LIC project staff, pretested at the LIC and Zhuhai Land Bureau, and modified according to the pretest results.

After the *key user* interviews, the project team incorporated the information and the comments from the interviews, and finalized the questionnaire format; meanwhile, the finalized questionnaire was translated by the project team into Chinese, and the questionnaire explanatory notes were prepared for the respondent organizations. During the task, the Chinese project staff also received training on questionnaire surveying and supporting techniques from the LIC's staff.

A questionnaire distribution list of twenty-six organizations was produced, and four responsible Chinese project staff have been assigned for tracking the questionnaires, answering inquiries, collecting and checking, and translating the responses into the forms prepared by the LIC staff for the data-modeling tools.

The questionnaires were distributed, by the LDGP and Zhuhai project staff, to participating organizations in December 1993. Participating organizations took 1 month to complete the questionnaire with the support from the LDGP and Zhuhai project staff, by answering inquiries, providing additional information where requested, and collecting the completed questionnaires within the prescribed time limit.

At the completion of the collection stage, LDGP and Zhuhai project personnel undertook a number of tasks:

- Quality assurance of the responses, including:
  - identification and resolution of missing data in response
  - identification and resolution of anomalous or contradictory data in response

- Translation of data from the questionnaire responses in Chinese to the standard format provided for data entry in English
- Transmission of the translated responses, on the basis of completed organizational responses— this effectively provided the LIC project team with the responses at the earliest possible time, providing the benefits of:
  - data being loaded into the CASE tool at the earliest possible moment
  - early detection of missing data and/or inconsistencies enhancing the ability to get early resolution
  - early feedback to the LDGP concerning the evolving state of the data model

Data modeling

Prioritize data elements

During this task, the LIC's project team analyzed the results of the questionnaire responses in order to determine the priorities of the identified data elements. This process produces the mandatory 'core data set' required for an implementation of the proposed system. In addition, those data items that are considered desirable or optional, are identified. The collection of this data is critical for the later stages of the project, in particular for the development of a data-capture program, but also, where required, for the identification of duplicate processes, and custodial and currency conflicts.

Create current physical model

This task is undertaken by the LIC's project team by using the translated questionnaires in the form of a series of preformatted forms that have been returned by the Chinese project personnel. The development of the current physical data model is performed with the aid of the selected CASE tools.

The physical data model documents, which are all data items currently in use by the participating organizations, whether map, paper, microfiche or computer based, include:

- Entities and their description including:
  - entity name
  - current source
  - systems using entity
  - currency of data
  - purpose of entity
- Attributes for each entity, and their description, including:
  - attribute name
  - attribute type
  - attribute length
  - purpose of attribute
- Keys, including:
  - primary keys and how they are determined
  - foreign keys and their derivation
- Data relationships

In addition to the physical data model documents are all processes and functions that have been identified during the response process. These processes are documented at the lowest possible level, including:

- Process and function description:
  - process or function name
  - process or function description/purpose

- process or function current method of carrying out
- relationship of function or process to other function or process
- data coming into function or process
- algorithms or other processing used by function or process
- data coming out of a function or process
- Process or function

Publish physical data model

The current physical data model created will be released to the respondent organizations as the results of the questionnaire survey and the analysis of the responses become available. During this task, these organizations will provide feedback to the project team, which will permit an assessment of the accuracy of the data to be made. In addition, the priorities assigned to the data constituting the data model can be assessed against other organizational needs to ensure that the planned data-capture program reflects the realistic requirements of the participating organizations.

Create new physical model

During this task, the physical model of the current systems will be transformed into a physical model of the new system. Several factors will influence this transformation process:

- existing work practices
- nature of transactions identified during the modeling of the functions and processes
- organizational responsibilities and custodial relationships
- target database software platform(s)
- target implementation software platform(s)

This physical model will be stored in the CASE tool, which will be used to report the new physical model. The model stored in the CASE tool may also be used at a later date for the semiautomated generation of a series of applications, which can be used to prototype the proposed applications.

Pilot system development and training program

According to the plan, the system development task will commence in September 1994 and will finish by early February 1995 in Australia. During this task the LDGP and Zhuhai will send four technical officers to participate in the system development as well as training for the program, specially set up for this project.

Hardware and software tools

The Zhuhai Pilot Project requires both a GIS and a RDBMS software for the project implementation, even though the project itself is vendor independent.

GenaMap and Oracle were selected for the project, due to their availability at the Land Information Centre. The LIC has existing licenses for these packages and these licenses can be used without extra cost to the project. In addition, both packages have NLS (National Language Support) options, in this case Chinese NLS. The platform is Hewlett Packard's HPUX and it also supports Chinese NLS.

Oracle CASE tools are used for data analysis, data modeling and system development, for which LIC has a license; it is not available to the Chinese market. The use of CASE tool can increase productivity by 40% to 70% over the traditional development methods and shorten the development process.

Pilot system development
The pilot system is a prioritized application subset of the future fully functional Land Information System for Zhuhai City. It will be selected by the project team from the data model by analyzing the importance of the applications to the whole system. It must be an 'open system' to be expanded to a fully functional LIS/GIS for Zhuhai City in the near future, and of course, have Chinese language capability.

The pilot system, once developed, will be tested and modified in Gongbei pilot area.

Training programs
The training issue is a key issue of the project, and the training programs are designed to be cost-effective to the participating organizations. So far, a number of training programs have been carried out.

Pilot system implementation
The *pilot system* implementation will commence after the pilot system development and on-site trial. During this task, a spatial database will be designed, and a data-capture program will be developed to suit Zhuhai government's needs, which later will be transferred into other areas in China.

Spatial database design and development
Existing data sets will be documented and related, and existing Chinese data standards and specifications will be collected, together with the results of the *user requirements study*, which is needed to design a spatial database and to define data-capture specifications. This process will be carried out by an experienced spatial database management specialist from the LIC, with assistance from Chinese colleagues. The spatial data specification will be published and reviewed by the LDGP and Zhuhai Land Bureau to ensure that the specification is suitable for Zhuhai and the Chinese environment.

Generate data-capture program
This task will concentrate on the development of a data-capture program, which will drive the acquisition of data for the proposed systems. The data-capture program will focus on a number of areas:
- mechanism of data capture, management and transfer
- hardware and software acquisition program
- quality assurance procedures
- specific data sets required by the LDGP, Zhuhai Land Bureau and other participating organizations

The thrust of the data-capture program will be to produce a plan for data migration (transfer), which is applicable not only in the Guangdong Province, but which can also be migrated into other provinces of China–in line with the project goals.

The identification of the data sets required to be migrated will be derived from the physical data model of new systems, and the specific data requirements will be fully documented in order that a quality assurance program can be implemented.

Produce final reports
All material produced during the project will be collated into a single document. This document will include:

- project plan
- quality assurance programs
- questionnaires and responses
- physical characteristics of current and new systems
- pilot system documentation
- data-capture program

In addition, a draft implementation plan will be generated, which will aid the LDGP and Zhuhai city authorities in determining the most appropriate strategy for implementing the proposed system. This implementation plan will be in a form suitable for transfer to other provinces in China.

Case Study 3

Australian Bureau of Statistics, Australia

GIS and a Spatial Information System for mapping a human resources profile of the country*

By John Mobbs, Project Manager, Land Information Centre, Department of Conservation and Land Management, University of New South Wales, Sydney

Background

The Australian Bureau of Statistics (ABS) is responsible for mapping the nation's human resources through the Census of Population and Housing. Spatial Information Systems (SISs) have provided the ability to add a powerful new perspective to this 5 yearly human resource profile, both in contributing to increased efficiency in the execution of the census, and in the data subsequently available to the ABS and ABS customers. The ABS has recognized the contribution that a SIS will make, and is committed to spend $5 million on Spatial Information Systems data and services in the lead-up to the next census in 1996.

The design and revision of an estimated 33 000 separate census collection maps within 145 census divisions is a labor-intensive and time-consuming task. The ABS recognized that the utilization of a SIS to automate these processes would contribute significantly to its long-term goal of saving public funds. Census collectors require paper maps for navigation and they use these maps to develop intimate knowledge of their collection districts. Districts vary in size, ranging from a city block area to a large area of the outback. This requires the flexibility to produce maps to a variety of scales, depending on the size of the collection district. The digital mapping functionality provided by an SIS is the only technology that can meet this need in a timely fashion.

* This paper was prepared by AISIST for inclusion in 'Australian Spatial Information Systems Industry Survey', 1994, a copy of which can be obtained from Price Waterhouse Urwick, 215 Spring Street, Melbourne 3000.

The ABS will utilize SISs to automate census district design processes and the production of census collection maps for field use. ABS analysts will also subsequently use the map base in collating data. Other applications will include digital dissemination products, such as CDATA on CD-ROM.

Project members

This SIS project is significant in two respects:

● There will be a digital map data collection for all settled areas throughout Australia. It is on a magnitude that has not previously been attempted in Australia.

● Every census collection map produced will be used in the field. This will allow for feedback that will be invaluable during the maintenance and upgrade phase of the project post-1996.

No single organization could undertake a project of this magnitude on its own, and the ABS is collaborating with both private and public sector organizations to achieve its goals. A Public Sector Mapping Agency (PSMA) consortium has been formed, and includes the following state, territory and federal departments:

● Federal: Department of Administrative Services/Australian Survey and Land Information Group (AUSLIG)

● Queensland: Department of Lands/Land Boundaries

● New South Wales: Department of Conservation and Land Management /Land Information Centre

Data issues

The primary purpose of the digital data is to facilitate the production of easy-to-read census collector maps over the differing landscapes of the entire Australian land mass. The data description therefore varies from what one might normally find on a conventional topographic map. A wide range of spatial data of different scale and theme will be used for the census project. Achieving the required level of detail, as well as consistency of data for nationwide mapping, is a significant challenge given that some states have been able to pursue their mapping responsibilities more comprehensively than others, due to their size, terrain type and population distribution, and, in some cases, due to their economy and resource allocation.

The data issues are more readily understood when one considers the aspects of scale and themes.

Scale

Data is available at varying scale and from different sources, dependent on location, which can be categorized as urban, rural or remote.

AUSLIG will be primarily responsible for digital datasets in remote zones. It will provide some 400 out of a total 530 map tiles from its digital product range, known as GEODATA TOPO-250K. Each tile is a topologically structured database suitable for GIS applications, and conforms to standard 1:250 000 map sheet boundaries. Other categories of data will need to be integrated within these tiles, for instance larger scale data to delineate the road and settlement patterns within towns and populated places.

The PSMA will source digital data for the rural zone from series mapping at scales between 1:25 000 and 1:100 000. Larger scale data will also be integrated over populated places.

In the urban zone the PSMA will use their Digital Cadastral Databases (DCDBs), supplemented with topographic data where appropriate. The source scales of plans used to create the various DCDB range between 1:2500 and 1:10 000.

Themes

The ABS project will use a wide range of themes for this project, all of which have particular significance for differing reasons.

Digital cadastral database

DCDBs are important for this project since the ABS intends to align census district boundaries with cadastral parcel boundaries, where no suitable topographic features exist. The cadastral line work for each state and territory will be supplied to the ABS for use as a DGN reference file in the MicroStation environment. In special cases the ABS may elect to print the cadastral line work on the census maps to assist census collectors with navigation in urban areas.

Local government areas

Local Government Area (LGA) boundaries are another important component of the data to be delivered. The states and territories are the custodians of these boundaries. The major statistical unit known as a census division is composed of one or more LGAs and many collected districts. Divisions, as well as census districts, therefore frequently share common boundaries with an LGA. The ABS staff require frequent access to accurate graphical representation of these boundaries in conjunction with topographic detail.

Relief

There is no relief layer in the PSMA data since contours frequently obscure other map data and are generally used only by experienced map users. Census collectors are frequently novices to the art of map reading, and usually travel by car. Interpretation of relief as portrayed by contours is, therefore, not necessary.

Buildings

Paradoxically for a census of population and housing, there will be no requirements to depict individual dwellings. This is because not all members of the PSMA collect to this level of detail, especially in urban areas. In addition, it is believed that the possible inclusion of uninhabited outbuildings with residential buildings would reduce the confidence and effectiveness of census collectors.

Vegetation

Vegetation is considered to be of little importance to census collectors and is thus omitted from the data specification.

Road network

Since most census collectors travel by motorized transport, the road access to dwellings is the most important PSMA data element and, indeed, is the single most valuable indicator of where dwellings exist. Emphasis is therefore placed on the production of road center lines.

In the urban zone and in populated places, these center lines are being produced from the DCDB. They will then be used by Candata, which will represent the street networks using the road casement style of presentation, found in street directories. This enhances the legibility and attractiveness of the map. Outside these zones, the center lines will be digitized from appropriate topographic map coverage. The spatial accuracy of the road center-line layer will therefore vary from 1 metre (1.09 yards) to 100 metres (109 yards).

Hydrography

A comprehensive representation of drainage features will be provided, because these are also an important source of navigational information, particularly in the rural and remote zones. The density of the drainage layer will vary according to the source of the digital data, since various compilation specifications will have been used on the original map source. The variations will be most evident at the transition between medium-scale and large-scale sources.

PSMA data description tables

Detailed tables have been prepared for each state and zone. This enables the ABS to identify the data sources and provides an estimate of the expected logical consistency of the data from different sources.

All digital map data will be integrated using MicroStation, and will be delivered in Intergraph design file format (DGN format).

Early lessons

One of the early lessons for PSMA staff, unused to comparing topographic and cadastral data, is the misalignment that commonly exists between the DCDB representation of roads and streams, and the topographic representation of the same features. Some misalignments are brought about by the scale differences between source documents, while others are due to the technology available at the time of compilation of the cadastral plans, that is ground survey versus modern and aerial photography. There is agreement within the PSMA that any such misalignments represent a future barrier to the adoption of the cadastre as the basis for LIS and GIS, and must be resolved as the project proceeds.

Future directions

In 1997 the ABS will conduct a review of the PSMA agreement and of the contribution of the data to the 1996 Census generally. This review will be followed by a maintenance and improvement phase. Data deficiencies will be addressed and enhancements will be effected, particularly with a view to implementing topologically correct GIS data structures for the state and territory datasets.

Possible enhancements and additions could include unique feature identifiers to allow for incremental feature updates, suburb boundaries, street addressing, postcodes, greater coverage at larger scales in the interior, inclusion of more cultural detail, precise alignment of topography to cadastre, and so on.

Technical overview of the ABS project data collection process

The ABS project will involve considerable volumes of data, with an estimated 57 gigabytes of disk storage being required. Coordination of such volumes will require competent project management skills, and the parties involved have drawn up a comprehensive program to master this task.

Figure 6.2 provides an overview of the process.

1. PSMA delivers data to Candata.
2. Candata carries out quality assurance (QA) on the data and loads it into the Intergraph-based mapping systems on their premises.
3. Candata aligns the 1991 census district (CD) boundaries with the map base data. In most cases these boundaries will be aligned to road center lines and topographic features, but in other cases the cadastral layer will have to be consulted where property boundaries have been used. This is common along LGA boundaries, which somewhere along their length are also a CD boundary. CD boundaries never cross LGA boundaries. Collection District Record Database (CDRD) records are married with spatial data and a tape is generated for transport to the ABS Canberra office.
4. The data is loaded onto an Intergraph 6750 UNIX-based server, which is located at the ABS's central computing facility. The data is integrated with the Master Spatial Database. This data is then viewed by the Central Office staff using a remote Intergraph 2730 UNIX machine, attached to NearNet, which is part of the ABS LAN. Following Central Office checks, the entire data package is transferred to the relevant State Office across the ABS network in off-peak hours.

5. The State Office receives the data on a local disk attached to an Intergraph Technical Desktop (TD1), which is running the census district (CD) design software that was developed for this project by *Candata* and *Navigate*. Sydney and Melbourne offices will each have two TD1 stations, with other State Offices each having one. TD1s are graphics optimized 486 DX66 configurations with sufficient disk space to hold the estimated data for the individual states or territories.

    Regional staff will use the CD design software based on an Oracle database, Intergraph's Microstation and Modular GIS Environment (MGE) Data Manager modules. They will update the CD boundaries to reflect anticipated population changes since 1991. As boundaries are moved in the redesign process, the associated CDRD records, being linked to the spatial data, are automatically moved to their new CD location, using the Oracle database on the centralized 6750 in Canberra.

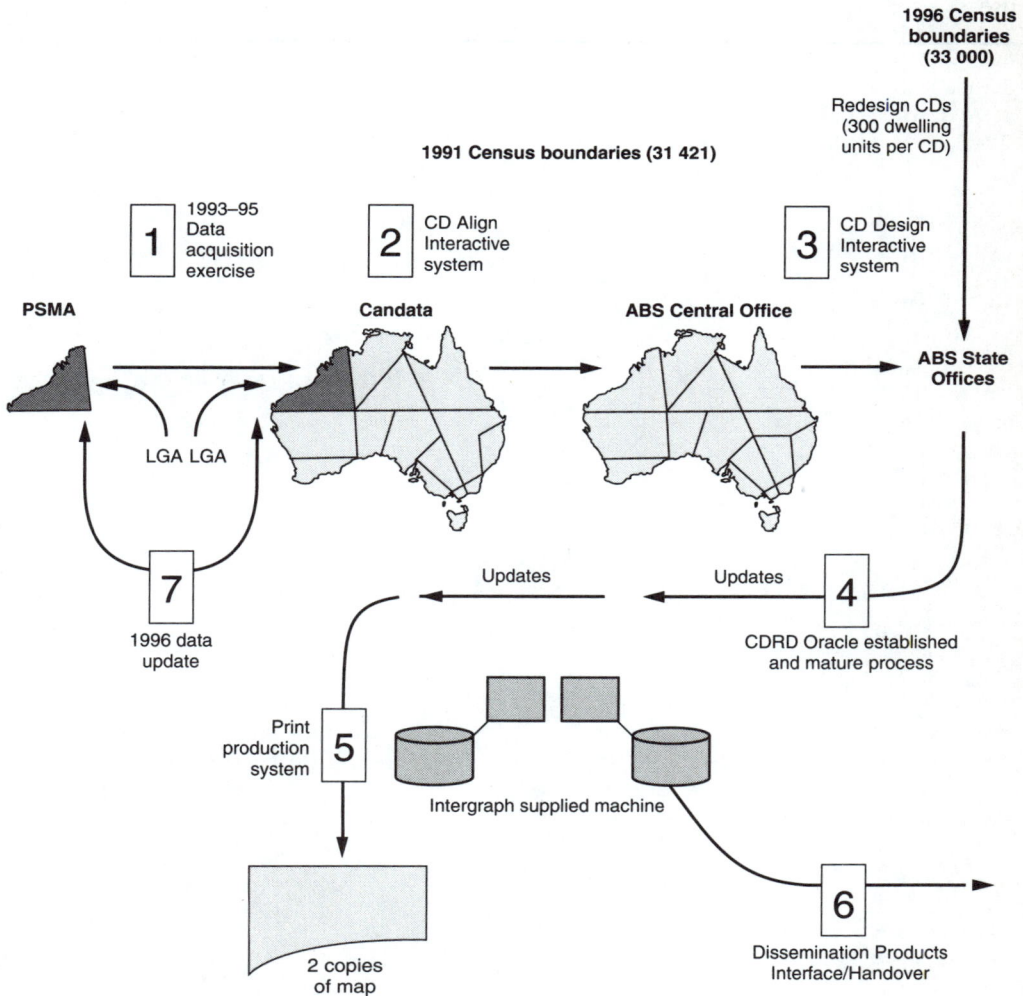

Fig. 6.2 *Data processing schema—1996 Census*

6. When this process is complete, the State Office staff will be looking at the 1996 census maps in the same format and detail as they will appear in the hands of the census collector, and will detect design errors immediately by reference to the amended CDRD. Changes are confirmed and sent again via the network to the 6750 in Canberra. At this stage both central and regional offices will view identical data.

7. The logic of the State Office design work is checked in Canberra and, when approved, the new CD boundaries and CDRD records are sent to Candata where these changes are incorporated into the mapping system. A plot file is produced for each CD. One CD will be printed per map, mostly on A4 or A3 sized paper. Two multicolor high-quality field managers will also be produced, displaying multiple census districts, which will be at smaller scales and on larger map sheets.

8. The whole Candata system will finally be migrated to the ABS, where it will reside for future use.

CHAPTER 7
Selected GIS/RS software

According to *International GIS Sourcebook* (GIS World 1993), there are nearly 300 different kinds of geographic information systems and related service software commercially available in the world. The real number could well be much higher than that since not all the vendors in the world participated in the survey. However, fully functioning GIS/RS software is far less than this figure. As GIS/RS technology matures, some will survive, but many of them are sure to lapse.

In this guideline, some the GIS software and RS software, based on functionality, are introduced, along with their applicability in the planning and decision-making environment.

There are basically two categories: vector-based GIS software and raster-based RS software. Most selected GIS software has some image-processing functionality, and RS software has some GIS capability, as well.

The selected GIS software is normally equipped with database management capability (or is able to run with commercial DBMS seamlessly), data/spatial analysis capability and digital mapping capability. RS software has all the capabilities of:

- image preprocessing
- image enhancement
- image extraction

The description of each software has the following content:

- a short description that highlights the characteristics of the software
- a reference price
- contact details for obtaining further information about the software

It is very difficult to specify the price of each software. Most software provides a core GIS software and several add-on or independent modules. The core GIS software may have a data management system (or interface to link with commercial DBMS dynamically), digital mapping capability and some basic spatial analysis capability. Such functions as image processing, network analysis, surface and other special functions are supplied separately.

Another factor is the discount policy. Normally, almost all vendors offer discount for any numbers of additional copies of the software, and a suite of the software package. Users from educational and research institutions, even governments, may get a better deal. Vendors may heavily cut back the price for market penetration.

Some vendors offer an exceptionally low price, even a free special package with full (or most) functions of the original products, for a limited time to allow users to have a try first.

Users sometimes request a vendor to temporarily install a system so that they can test alternative systems to identify which one suits their purpose.

7.1 GenaMap, with Genasys II

Genasys is an international developer, and a supplier of geographic information systems and associated products and services. With over 3500 systems installed, a network of office and partners throughout the world, and more than 15 years experience in GIS, Genasys leads the industry in innovative, *integrated spatial information systems* technologies. The flagship of the Genasys suite of spatial systems is the GIS product, GenaMap. Complementary modules to the GIS provide for 'Cell' modeling, 'Tin' modeling, enhanced analysis, networking, links to external RDBMS in a true client server environment, image visualization, document storage and retrieval, and many more extension services.

Genasys is provided with a database suitable for operation 'stand alone'. Where users are part of an enterprise system they can link the system to an external DBMS in true *client server* mode, and then store and retrieve their data externally. Multiple databases can be linked simultaneously and can be of different suppliers, for example Oracle, Ingres and Informix.

Genasys provides a module, GenaVive, which has a wide range of functions for handling and viewing preclassified remote-sensed images or scanned maps, orthophotos, diagrams, or anything else which can contribute to the richness of information able to be stored and analyzed in conjunction with vector data. Data formats range from monochrome and grayscale images to 8 bit and 24 bit color imagery. Input to GenaVive can be from a wide range of systems, including some of the more popular, such as ER Mapper, ERDAS and MicroBrian.

Genius, the Graphic User Interface to all products, which takes full advantage of the client server architecture, is event-driven, object-oriented and fully Motif 1.2 compliant.

Genius runs in two modes: starter or advanced. In advanced mode the user has access to a comprehensive tool set as well as numerous application component templates to facilitate application development.

The GIS is complemented by four civil engineering modules for surveying, terrain modeling, road design and hydrology.

Reference price:

Single license Workstation: A$25 350. Annual fee: A$3802.5

Single license PC: A$20 280.
Annual fee: A$3402.00

5 user network license: A$97 175.
Annual fee: A$14 576.25

GenaCell, GenaTin and GenaVive are each 30% of GenaMap pricing for specified user level.
GenaRave is available separately at US$10 000 per copy.

Contact details
Genasys SDN BHD
Suite 1101, 11th Floor, Plaza Atrium
(Attention: Mr Lorong P. Ramlee)
50250 Kuala Lumpur, Malaysia
Phone: (60-3) 238 6403
Fax: (60-3) 238 5640

Genasys China
Genasys China
Xiyuan Hotel Suite 10801
#1 Sanlihe Rd
Beijing, China 100046
Phone: (86-1) 8313388 10801
Fax: (81-1) 8313388 10841

7.2 TNTmips, with Microimages Inc.

TNTmips (Map and Image Processing System) is an integrated image processing, GIS, CAD, spatial database management and desktop mapping system, used in sixty nations from

Angola to Zimbabwe. It is widely used in geologic mapping and analysis; the combining of satellite images, map overlays and geolocated database materials into maps; land management; environmental issue analysis; and the organization, storage and analysis of massive amounts of spatial materials.

TNTatlas provides a unique, low-cost innovative approach for the publication and distribution of spatial information via CD-ROM or network. Massive printed records, database and other materials can be economically and rapidly disseminated for immediate access on all computer types, including portable. Raster, vector, CAD, text or database materials can be combined in TNTatlas.

TNTview is an interactive tool for the complex combination, visualization and interpretation of large raster, vector, CAD, geolocated database and text example. Start with several image layers, which are automatically mosaiced during display, and then overlay linear map features to which are attached attributes stored in a relational database.

TNTsdk for TNTmips is a software development kit that allows one to create custom processes for use with a TNTmips. TNTsdk is purchased with an initial license fee and can be upgraded quarterly by annual subscription. TNTsdk is designed for use with TNTmips only.

TNTlink for TNTmips is used with TNTmips to prepare Hyper-index stacks from spatially interrelated datasets made up of raster, vector and CAD layers, with associated relational database tables and text.

TNTdemo is a 30-day trial package that includes CD-ROM(s) from which to install all the TNT products for every platform supported by MicroImages; a 300-page illustrated primer entitled *Guide to Map and Image Processing*; the Application Note series, currently nine monographs with 1200 pages of illustrated text; TNTmips On-line Documentation providing the equivalent of 1600 pages of illustrated text; the TNT products *Detailed Installation Manual*; a physical authorization key that is attached to the serial port of any

supported computer platform, and a full US$200 refund upon purchase of any new TNT product and return of the physical authorization key.

Platforms supported are:

- IBM PC and compatible (DOS, Windows, and Window NT)
- Apple Macintosh (System 7)
- Sun SPARC station (Solaris)
- DEC Alpha AXP (NT or OSF.1)
- DECstation (Ultrix)
- IBM RS/6000 (AIX)
- HP Apollo 700 series (HP-UX)
- Data General Aviion via (DG-UX)

Reference price
TNTmips from US$5000 for PC, to US$10 000 for DEC/3 & 5000.

TNTsdk: from US$300 to US$500
TNTlink: US$1000

TNTview: US$1000–US$1100
TNTatlas: US$600

Annual fee: from US$800 for PC to US$3300 for Unix workstation

Contact details
Taiwan

New Image Co. Ltd
Attention: Mr Gregory Y. Tang
5F, N. 179 Shui Yuan Road
Taipei, Taiwan
Phone: (886) 230 39836
Fax: (886) 230 99120

Malaysia

Cenanet (SEA) Sdn Bhd
(Attention: Mr Alfred Yeap)
4, Jalan SS21/62
Damansara Utama
47400 Petaling Jaya, Selangor
West Malaysia
Phone: (60-3) 719 0670
Fax: (60-3) 719 0671

Thailand

Earth Intelligence Tech. Co., Ltd
(Attention: Boonchoob Boongthong)
203/3-4 Moo 6
Pattanakarn Road
Prawet, Bangkok 10250
Thailand
Phone: (66-2) 321 74312
Fax: (66-2) 321 4338

Japan

OpenGIS Corporation
(Attention: Toshihiko Waza)
1-19-14 Azuma-bashi
Shumdia-hu
Tokyo 130
Japan
Phone: (81-3) 362 32851
Fax: (81-3) 362 33025

Iran

Chkon Co. North Eskandari St
(Attention: Sohrab Batmanglidj)
Mohamadi Alley, #21
Tehran 14197
Iran
Phone: (98-2) 192 6411
Fax: (98-2) 192 7908

7.3 MicroStation and MGE, with Intergraph

Intergraph is a *Fortune 500* company with offices located in forty-two countries. With 25 years experience in the interactive computer graphics industry, Intergraph is dedicated to providing high-quality hardware, software and services that require superior computer graphics technologies.

Intergraph provides mapping/GIS solutions for government infrastructure, environmental and natural resource management and digital cartography, based in Intergraph's MGE (Modular GIS Environment). MGE's comprehensive suite of modules for project and data management, data collection and integration,

spatial query and analysis, and output support GIS and mapping work flows in every major industry.

The MGE platform serves as an integration point for all Intergraph GIS and mapping applications. MGE is built on MicroStation, a widely used CAD/CAM software, and a Relational Interface System (RIS) to combine superior graphic display and manipulation capabilities with attribution storage in an industry standard relational database management system (RDBMS). The RIS provides a generic interface to most RDBMSs. An RIS offers simultaneous, direct and real-time read/write connections to local or remote RDBMSs from many vendors which are on dissimilar hardware platforms using numerous protocols. An RIS makes an entire network of databases available as if they were a single, local database.

Intergraph supports image processing within the spatial context of a complete geodetically controlled environment. The Intergraph image-processing solutions are integrated with the MGE products to provide a truly seamless link between imaging and the GIS. The complete range of Intergraph image-processing products includes soft-copy photogrammetry capabilities through to raster vector integration for high-quality map production.

Application modules that are based on the MGE platform include:

- *MGE GeoData Manager*
  Feature level data management environment for a seamless map base
- *MGE Parcel Manager*
  Client application for maintenance of cadastral parcel fabric
- *MGE Projection Manager*
  Complete suite of geometric transformations to support common map projections/coordinate systems
- *MGE Analyst*
  Spatial analysis and display of topologically structured geographic data
- *MGE Dynamic Analysis*
  Dynamic query interface for object-based analysis of MGE data sets

- *MGE Grid Analyst*
  For manipulation, displaying and analyzing multiple layers of grid data
- *MGE Terrain Modeler*
  Surface analysis of contours and drainage data using triangle or grid based terrain models
- *MGE Network Modeler*
  For modeling and management of utility networks in a GIS environment
- *MGE Segment Manager*
  Dynamic segmentation of linear graphics, providing the ability to display and analyze any portion of a linear feature without modifying the base map
- *MGE Voxel Analyst*
  A general purpose visualization tool for dynamic modeling of geotechnical data in the context of its geographic location
- *MGE Environment Manager*
  A comprehensive set of utilities for the management, input, verification and presentation of environmental data

Reference price

MicroStation: A$6000
Basic MGE: A$3000–$10 000

Contact details

China

Intergraph China Shanghai
Liaison Office
1505 Jing An Guest House
370 Hua Shan Road
Shanghai, 200040
Phone: (86-21) 255 1888
Fax: (86-21) 255 2657

Hong Kong

Intergraph Graphics System Hong Kong Limited
3606-10 China Resources Building
26 Harbour Road
Wan Chai, Hong Kong
Phone: (852) 593 1500
Fax: (852) 802 0781

Japan

Intergraph Japan K.K
Nikko Motoyoyogi Building 2F
30–13 Motoyoyogi-Cho Shibuya-Ku
151 Japan
Phone: (81-3) 5453 3011
Fax: (81-3) 55453 3038

Korea (South)

Intergraph Korea, Ltd
73-79 Yongho-Dong
Kyungsangnam-Do
Changwon 641-041, Korea
Phone: (82-2) 521 7102
Fax: (82-2) 521 7109

Singapore

Intergraph Systems South-East Asia (PTE) Ltd
Block 85 Science Park Drive
#02-01/04 The Cavendish 0511
Phone: (65) 779 0177

Taiwan

Intergraph Corporation Taiwan
11th Floor, 21st Century Building 207
Tung Hwa N. Road, Taipei 10592
Taiwan
Phone: (886-2) 514 9282
Fax: (886 2) 514 9303

7.4 SPANS, with TYDAC Technologies Inc.

TYDAC Technologies Inc., established in 1982, evolved from a group of professionals committed to the development of applications-oriented, versatile and cost-effective analytical spatial solutions.

The SPANS (SPatial ANalysis System) product line consists of a range of spatial analysis products and services developed by TYDAC. The company's two main products are SPANS MAP, a data visualization and query tool, and SPANS GIS, a complete spatial analysis and modeling system. To date, with the support of a worldwide network of business partners, more than 3500 SPANS systems have been installed in sixty-four countries across the globe.

SPANS is a modular system, with products available for the Windows, OS/2 and UNIX operating system environments. The complete suite of software modules is the SPANS GIS system. The standard package is GIS Builder, a set of basic tools for building databases, constructing analytical models and visualizing and querying data.

Additional modules are available to expand the functions of GIS Builder. The flexibility allows you to tailor SPANS to your specific requirements and applications. The following SPANS modules are all stand-alone, in that they require no other modules to run:

- SPANS MAP
  Desktop mapping package
- SPANS TYDIG
  Digitizing package
- Data Translation
  Data translation modules (raster and vector interfaces)
- GIS Builder I
  Basic spatial analysis functions (includes TYDIG)
- GIS Builder II
  Basic spatial analysis functions (includes TYDIG and Data Translation)

The analytical capabilities of SPANS analysis model can be grouped into four major classes: transform, analysis, identify and model.

The following analytical modules can be fully integrated into GIS Builder I or II:

- 3-D View
- Contouring
- Surface Generator
- Point Aggregation
- Neighborhood analysis
- Network Analysis
- Topological Analysis
- Visibility Analysis
- Interaction Modeling
- Multicriteria Modeling
- Map Modeling
- Table Modeling

- Point Modeling
- Application Developer Program

Reference price

SPANS GIS Builder I: A$5900
SPANS GIS Builder II: A$8900
SPANS MAP: A$999
Full price (PC): A$26 500; Workstation: A$29 500

Contact details

Indonesia

P.T. Indra Bisma Mahesa
Jl. Raden Salch No 4
Jakarta, 10430, Indonesia
Phone: (62-21) 390 3884
Fax: (62-21) 310 6295

Japan

CRC Research Institute Inc.
Information System Division
D18, 1-3, Nakase, Mihama-Ku
Chiba City 261-01, Japan
Fax: (81-43) 733 2145

Korea (South)

Hyundai Electronics Industries Co Ltd
9th Floor, Hyundai Jeonja Bldg.
66, Jeokseon-Dong, Chongro-Ku,
Seoul 110-062, Korea
Fax: (82-2) 733 2145

Malaysia

Mesiniaga Sdn. Bhd
Menara Mesiniaga
1A, Jalan 55 16/1, Subang Jaya
47500 Petaling Jaya
Selangor Darul Ehsan, Malaysia
Fax: (60-3) 736 3838

Philippines

Project Consultants Group Inc.
P.B. Dionisio Building, Suite 317
27 Don Alejandro Ave
Quezon City, Philippines
Fax: (63-2) 922 2882

Thailand

Earth Intelligence Technologies Co. Ltd
203/3-4 Moo 5, Pattanakarn Road
Prawet, Bangkok 10250
Thailand
Fax: (66-2) 321 4338

7.5 ARC/INFO with Environmental Systems Research Institute

All Environmental Systems Research Institute (ESRI) software products are designed around a single, integrated data model. This enables each ARC/INFO software component to seamlessly interact with the others. This emphasis on data interconnections enables ARC/INFO to adapt to stringent GIS requirements.

ARC/INFO, ESRI's flagship software product, is a high-end GIS that consists of a complete geoprocessing toolbox for the automation, modification, management, analysis and display of geographic information. Core functionality includes:

- GIS graphical user interface: (ArcTools)
- Map projections, manipulation and management: (ARC)
- Polygon overlay and buffering: (ARCPLOT)
- Spatial data analysis: (ARCPLOT)
- Data conversion: (ARC)
- Data entry and edit tools: (ARCEDIT)
- Data display and analysis tools: (ARCPLOT)
- Cartographic publishing tools: (ARCPLOT)
- Plotter support: (ARCPLOT)
- Geographic data management: (LIBRARIAN)
- Macro language: (AML)
- RDBMS Access: (DATABASE INTEGRATOR)
- Tabular data management: (INFO)
- Image display and management: (IMAGE INTEGRATOR)

- Linear modeling: (Dynamic Segmentation)
- Overlapping polygons and area modeling: (Regions)
- Interactive graphic form menu creation: (FormEdit)

Optional extension products available to further expand the capabilities of ARC/INFO for specific applications are:

- TIN: the ARC/INFO surface modeling extensions
- GRID: the ARC/INFO raster modeling extensions
- NETWORK: the ARC/INFO network analysis extensions
- COGO: the ARC/INFO coordinate geometry extensions
- ArcScan: the scanning data entry extensions
- ArcStorm: advanced spatial data management extensions
- ArcExpress: display speed enhancement extensions

The software can run on UNIX and Open VMS workstations from DEC, IBM, Sun, HP, NEC, Data General and Silicon Graphics.

Reference price

Contact distributor for ARC/INFO pricing.

Contact details

China

Super Full Technology Co. Ltd
(Attention: Mr Francis Ho)
Unit 1202, Landmark Building,
8 North Dongsanhuan Road
Chaoyang District, Beijing 100004,
China
Phone: (81-1) 501 2136
Fax: (86-1) 501 2137

Hong Kong

The GMR High Technology Ltd
(Attention: Mr Danny Cheung)
Cosmos Building, Room 403-5
8-11 Lan Kwai Fong, Central, Hong Kong

Phone: (852) 523 8668
Fax: (852) 525 6622

India

National Institute of Information
Technologies Ltd
Attention: Mr Sundeep Srivastava
NIIT House
C-125, Okhla Phase I
New Delhi 110 020, India
Phone: (91-11) 6812 7341-3
Fax: (91-11) 681 7344

Japan

PASCO Corporation
(Attention: Mr T. Okuyama)
No. 13-5, 2-Chome, Higashiyama
Meguro-ku, Tokyo 153, Japan
Phone: (81-3) 3715 1601
Fax: (81-3) 3715 1607

Korea (South)

Cadland (Seoul Mapping Research Institute)
Attention: Mr Jay J Yoon
Songuam Bldg, Suite 312, 1358-6,
Seocho 2-dong,
Seocho-gu, Seoul, Korea
Phone: (82-2) 557 4888
Fax: (82-2) 554 2096

Malaysia

ESRI-South Asia
Environmental Systems (M) Sdn. Bhd.
(Attention: Mr Richard Teh)
1A Wisma Siong Huant
13 Jalan 223, 46100 Petaling Jaya
Selangor, Malaysia
Phone: (60-3) 757 9930
Fax: (60-3) 757 9679

Pakistan

Pakistan Resources Development Services, Ltd
(Attention: Mr Naved Zaheer)
29, Block 7 & 8 D.C.H.S.
Sharea Faisal
Karachi-75350, Pakistan

Phone: (92-21) 453 0926
Fax: (92-21) 453 4619

Singapore

ESRI-South Asia
(Attention: Mr Jim Durana)
350 Orchard Road
Shaw House, #12-01/03
Singapore 0923, Singapore
Phone: (65) 735 8755
Fax: (65) 735 5629

Thailand

ESRI-Thailand
Attention: Mr Krairop Luang-Uthai
U Chuliang Foundation Bldg
1st Floor
968 Rama IV Road
Bangkok 10500, Thailand
Phone: (66-2) 233 7811
Fax: (66-2) 236 6953

7.6 Smallworld GIS, with Smallworld Systems Ltd

The Smallworld GIS product is based upon the principle that successful systems are built on very strong foundations. General features include:

- high-level customization environment
- user interface environment
- seamless mapbase
- data integrity management
- multiuser version management
- external DBMS integration
- standards
- toolkit of standard utilities
- interfaces to hardware and other software systems

Technical features include:

- Object-oriented software environment provides:
  - the powerful object-oriented language MAGIK to develop complex one-off customizations

- the CASE tool to allow the application data model, including its topology, to be described and the user interface to be laid out
- the application configuration environment where the GIS environment is tailored in different ways to meet the user interface needs of different types of users
- Integrated raster/vector/object database
- Topologically structured geometry—a variety of generic applications are provided, which are used as the basis of building specific applications, such as:
  - network tracing
  - polygon operations
  - query system
  - object editor
  - topology editor
  - geometric constructions
  - graphical styles
  - real world data modeler
  - report generator
  - on-screen digitizing

Reference price:

Single user, minimal function: A$5000
Single user, full function: A$50 000
Five user, full function: A$15 000

Contact details

UK

Smallworld Systems Ltd
Burleigh House, 13-15 Newmarket Road
Cambridge, CB5 8EG, UK
Phone: (44-223) 460 199
Fax: (44-223) 460 210

Australia

Level 19, Monash House
15 William Street
Melbourne, VIC 3001, Australia
Phone: (61-3) 691 4217
Fax: (61-3) 691 4212

7.7 MapInfo, with MapInfo Corp.

MapInfo is basically an ease-of-use, customizable and affordable GIS software. Features include thematic mapping, geocoding, fully functional relational database with SQL capabilities and spatial extensions, vector and raster images display, buffer generation, overlay, project transformation, features drawing and data presentation.

The development environment, supported by MapBasic, can customize the interface, menu and icon to meet the requirement of various applications. MapInfo also provides the information on the third-part modules that can run within MapInfo.

MapInfo does not use topological structured vector data. It does not have the surface analysis, network analysis and sophisticated image-processing capabilities.

Reference price

MapInfo for Windows (PC): A$2350
MapInfo for Mac: A$2350.
MapInfo for Sun, HP (Unix): A$4375

Contact details

Australia

Level 4, 170 Pacific Highway,
Greenwich NSW 2065
Phone: (61-2) 437 6255
Fax: (61-2) 439 1773

Malaysia

24A Lorong 4/50
Petaling Jaya
Selangor Darul Ehsan, 46050 Malaysia
Phone: (60-3) 793 5373
Fax: (60-3) 793 5378

Papua New Guinea

Theodist
Spring Gardens Road
Port Moresby
Papua New Guinea
Phone: (675) 256 500
Fax: (675) 250 302

7.8 Mapper, with Earth Resources Mapping Pty Ltd

ER Mapper is the premier image-processing system used to process satellite, geophysical, seismic and airborne data. Applications include forestry, land information, mineral exploration, oil and gas, surveying and water resources.

The Graphic User Interface is easy to learn and is friendly to use. It includes complete on-line help and manuals. Each copy of ER Mapper includes sixty example datasets, and a complete suite of 200 processing algorithms that show you how to process earth science data for a wide range of applications.

A wide range of raster and vector data may be processed, mosaiced, combined, analyzed and integrated with information stored in GIS and DBMS systems. ER Mapper also provides a breakthrough—true interactive image processing with the innovative dynamic algorithm compiler.

ER Mapper (Version 5.0) features includes:

- image viewer
- map composition
- hardcopy map generation
- 243 hardcopy formats
- PostScript engine
- 129 import/exports
- edit of ARC/INFO coverages
- GenaMap link
- vector annotation
- dynamic algorithms
- 3-D perspective
- 3-D flythrough
- 3-D stereo image hardcopy
- C developers toolkit
- image interpretation
- algorithm spatial modeler
- digitizer support
- image rectification
- stats and scattergrams
- raster to vector
- ISOCLASS and supervised classification
- batch engine
- toolbars
- FFT processing
- dynamic data fusion
- dynamic cell fusion
- traverse extraction
- background processing
- geolinked windows
- map projection database
- standard filters and formula library

New features, which will be available with ER Mapper 5.1 in early 1995, include:

- DTM generation from stereo pairs
- DTM editing in 3-D stereo
- image orthorectification
- feature editing in 3-D stereo

Reference price
Commercial

ER Mapper 5.0/UNI, one floating license on a network: US$19 500
ER Mapper 5.0/PC: US$6500.

Educational

ER Mapper 5.0/Unix or PC, five licenses: US$2000.

Contact details
China

Tradeglobe Import & Export
Beijing, China
Phone: (86-1) 506 6560
Fax: (86-1) 506 6563

Hong Kong

Leica Instruments
Quarry Bay, Hong Kong
Phone: (852) 564 2299
Fax: (852) 564 4163

India

KLG Consultants Pty Ltd
New Delhi, India
Phone: (91-11) 573 4604
Fax: (91-11) 5575 5491

Indonesia

Citrodata Intersystem PT
Jakarta, Indonesia
Phone: (62-21) 489 6883
Fax: (62-21) 489 8881

Japan

Image & Measurement Inc.
Tokyo, Japan
Phone: (81-3) 3365 3641
Fax: (81-3) 3365 3646

Korea (South)

Leica Instruments
Seoul, Korea
Phone: (82-2) 514 6543
Fax: (82-2) 514 6548

Malaysia

Malaysian Business Advisors
Kuala Lumpur, Malaysia
Phone: (60-3) 443 8005
Fax: (60-3) 443 9410

Singapore

Landmark Graphics International Inc.
Singapore
Phone: (65) 338 5833
Fax: (65) 338 9228

Taiwan

Quantitative Technology
Taipei
Phone: (886-2) 737 2105
Fax: (886-2) 736 7210

Thailand

ACS
Bangkok, Thailand
Phone: (66-2) 439 5791
Fax: (66-2) 437 4422

7.9 IDRISI, with Clark University

IDRISI offers an extensive set of tools for database management and query, geographic analysis and statistical analysis. Database query tools include those for reclassification, overlay, cross-tabulation and area-perimeter measurements. Map algebra, distance analysis, surface generation and characterization, and neighborhood analysis tools are also included. Spatial and attribute data may be stored separately, with a link to a spatial definition image file.

IDRISI is designed to be easy to use, yet provide highly professional GIS, image processing and spatial statistics analytical capability. No expensive graphic cards or peripheral devices are required to take full advantage of the analytical power of the system. IDRISI's fully documented and open architecture, along with its question-and-answer interface, make learning IDRISI quite easy. In addition to the standard GIS capabilities offered, the IDRISI package also includes tools for advanced analyses and modeling such as, multicriteria/multiobjective decision making, Fuzzy Set analysis, bayesian uncertainty analysis and anisotropic cost-surface generation, to name but a few. The advantage of IDRISI is that if offers such a broad range of capabilities, yet it is inexpensive, easy to learn and to use, and it operates on the PC platform.

IDIRIS offers a full suite of image-processing capabilities including procedures for georeferencing (rubber-sheet and projection) and image restoration, image enhancement (for example, filtering and contrast manipulation), image transformation (for example principal components, RGB/HLS), and image classification, including supervised (Parallel-piped, Minimum distance to Means and Maximum Likelihood classifiers) and unsupervised techniques.

Reference price

Commercial/Private: US$640
No-Profit/Research/Government: US$320
Full-time student: US$160
Site license (10 or more seats): US$160 each

Contact details

The sole distributor:
IDIRIS Project
Clark University
950 Main Street Worcester
Massachusetts 01610-1477
Phone: (1-508) 793 7526
Fax: (1-508) 793 8842

References

AISIST 1993a, *AISIST Training Modules*, Unit 2.03.01, Australian Institute of Spatial Information Sciences and Technology, Australia, Land Information Centre, Bathurst, NSW.

AISIST 1993b, *Basic Data Types, Integration & Linkage*, AISIST Training Modules, Unit 2.0101, Australian Institute of Spatial Information Sciences and Technology, Land Information Centre, Bathurst, NSW.

AISIST 1993c, *What is GIS? Why GIS?*, AISIST Training Modules, Unit 1.0101, Australian Institute of Spatial Information Sciences and Technology, Land Information Centre, Bathurst, NSW.

Behrsin, M., Mason, G. & Sharpe, T. 1994, *Reshaping IT for Business Flexibility—The IT Architecture as a Common Language for Dealing with Change*, IBM Series, McGraw-Hill Inc., Europe.

Berry, J. K. 1987, 'Fundamental operations in computer-assisted map analysis', *International Journal of Geographic Information Systems*, vol. 1, no. 2, pp. 119–36.

Brisbane City Council 1994, Bitmap Access System Product Catalogue, unpublished.

Bryson, J. M. 1993, *The Strategic Planning Process: Strategic Planning for Public Service and Non-Profit Organisations*, Pergamon Press, Oxford.

Carter, J. 1988, *On defining the Geographic Information System, Fundamentals of Geographic Information Systems: A Compendium*, American Society for Photogrammetry and Remote Sensing and American Congress of Surveying and Mapping, Falls Church, Virginia, pp. 3–8.

Cartwright, J. C. 1993, *CD-ROM feeds GIS data appetite*, GIS World, Fort Collins, Colorado, Oct.

Castle, G. H. 1993, *Profiting from a Geographic Information System*, GIS World, Fort Collins, Colorado.

Coleman, D. & McLoughlin, J. 1994, 'Building a global spatial data base infrastructure: usage paradigms and market influences', *Geomatica*, vol. 48, no. 3, pp. 225–36.

Curran, P. J. 1985, *Principles of Remote Sensing*, Longman, London.

Deuker, J. J. 1979, 'Land resource information system: Spatial and attribute resolution issues', *Proceedings, International Symposium on Cartography and Computing: Auto-Carto IV*, vol. 2, pp. 328–36. *Dictionary of Computing* 1990, 3rd edn, Oxford University Press, Oxford.

ESCAP 1990, *State of the Environment in Asia and the Pacific*, Economic and Social Commission for Asia and Pacific, Bangkok, Thailand.

Garner, B. J. & Zhou, Q. 1990, *GIS and RS—Towards Better Integration of Data for Land Resource Management*, AURISA '90, URPIS '18, Canberra, Nov., pp. 185–94.

Gatrell, A. C. 1990, *Geographical Information Systems*, vol. 1, *Concepts of Space and Geographical Data*, Longman Scientific and Technical, Harlow, UK.

GIS World 1989, *Spatial Data Exchange Formats*, GIS SourceBook, GIS World, Fort Collins, Colorado.

GIS World 1993, *PCs vs. workstations—Looking beyond the hype*, GIS World, Fort Collins, Colorado, June pp. 48–56.

Heit, M. & Shortreid, A. 1993, *GIS Applications in Natural Resources*, GIS World, Fort Collins, Colorado.

Huxhold W. E. 1991, *An Introduction to Urban Geographic Information Systems*, Oxford University Press, New York.

Johnson, R. W. & Granger, K. J. 1994, 'Hazard Management: Better Information for 21st Century', AURISA Workshop Emergency Management, AURISA, Canberra, ACT, April.

Lebreton, P. P. & Henning, D. A. 1961, *Planning Theory*, Prentice-Hall, Englewood Cliffs, New Jersey, p. 7.

LIC (Land Information Centre of NSW) 1993, *Computer Systems User Manual*, vol. 1, LIC, Bathurst, NSW.

Lyons, K. 1993, 'Data Quality Assurance and Control, A Producer Point of View', *Proceedings of Conferences on Land Information Management Geographic Information Systems Advanced Remote Sensing*, Sydney, NSW.

Mahoney, R. P. 1993, '*Best Practice Guidelines for GIS Implementation in GB Local Government*', World of GIS, AGI 93, 16–18 Nov. 1993, International Conference Centre, Birmingham.

Marble, D. F. & Amundson, S. E. 1988, 'Microcomputer based geographical information systems and their Role in urban and regional planning', *Environment and Planning*, B. 15, pp. 305–24.

Montgomery, G. E. 1993, *National Map Accuracy Standards*, GIS World, Fort Collins, Colorado.

Montgomery, G. E. & Schuch, H. E. 1993, *GIS Data Conversion Handbook*, GIS World, Fort Collins, Colorado.

Newell, R. G. 1994, The why and the how of the long transaction, *Geographic Information, 1994, The Source book for GIS*, AGI, Taylor & Francis, London.

NSW Government 1991, *Acquisition of Information Technology, Conceptual Framework*, New South Wales (Australia), Premier's Department, Office of Public Management, Sydney, NSW.

Oxborrow, E. 1986, *Database and Database Systems*, Chartwell-Bratt, Lund, Sweden.

Parent, P. & Church, R. 1987, 'Evolution of Geographic Information Systems as Decision Making Tools', American Society for Photogrammetry and Remote Sensing and American Congress on Surveying and Mapping, *Proceedings of GIS '87*, Fort Collins, Colorado, pp. 63–70.

Pearce, N. 1990, 'Taking the risk out of system selection', *Mapping Awareness*, vol. 4, no. 1, pp. 3–6.

Peuquet, D. J. 1977, *Raster Data Handling in Geographic Information Systems*, Geographic Information Systems Laboratory, State University of New York, Buffalo, New York.

Rajan, M. S. 1991, *Remote Sensing and Geographic Information System for Natural Resource Management*, Asian Development Bank, Manila, Philippines.

Ryan B. 1994, 'RISC grown up', *Australian Personal Computers*, Feb.

Salomonsson, O. 1980, 'Data gaps in the planning process: An application to environmental Planning', *Proceedings, Workshop on Information Requirements for Development Planning in Developing Countries*, International Institute for Aerospace Survey and Earth Sciences (ITC), Enschede, Netherlands.

Shelly, G. B. 1991, *System Analysis and Design*, Boyd & Fraser Publishing Company, San Francisco.

Skidmore, A. 1993, *Remote Sensing and GIS Working Group Report*, NSW Remote Sensing Committee, NSW Information Office, Sydney, NSW.

Smith, T. R. 1987, 'Requirements and principles for the implementation and construction of large-scale geographic information systems', *International Journal of Geographical Information Systems*, vol. 1, pp. 13–31.

Somogyi, E. & Gallier, R. D. 1987, *From Data Processing to Strategic Information Systems—A Historical Perspective*, Towards Strategic Information Systems, vol. 1, Tunbridge Wells, Kent.

Soursa, E. 1993, 'Development of a Pricing Policy for Geographic Information', *URISA Proceedings*, URISA, Washington.

Strand, E. J. 1993, *GIS Hardware Gets Better, Faster, and Cheaper*, GIS World, Fort Collins, Colorado, Oct.

Tanenbaum, A. S. 1990, *Structured Computer Organisation*, Prentice-Hall International, Inc., Englewood Cliffs, New Jersey.

Van der Lans, R. T. 1988, *Introduction to SQL*, Addison-Wesley Publishing Co., Reading, Massachusetts.

Yeh, A. C. 1991, 'The development and applications of geographic information system for urban and regional planning in developing countries, *International Journal of GIS*, vol. 5, no. 1, pp. 5–27.

Index

A

accident analysis, 59
Alberts Conic Equal-Area projection, 26
allocation
  demand, 32
  supply, 32
  impedance of supply, 32
Alper Records, 54
analog map document, 6
annotation, 34
ANSI (American National Standards Institute) standard, 95
application modules, 48
application of computers, 38
application system development, 19
applications, 38
  access from common data to networks of workstations, 38
  generation of geometrically corrected orthophotos and satellite images from raw data, 38
  integrating image-processing raster-based data with GIS vector-based data, 38
  mosaicing images together, 38
  multiple applications at the same time, 38
architecture of the hardware, 35
Arc/Info, 47, 54–5, 57–61, 129
  advanced spatial data management extensions: (ArcStorm), 129
  Arc/Info coordinate geometry extensions: (COGO), 129
  Arc/Info network analysis extensions: (NETWORK), 129
  Arc/Info raster modeling extensions: (GRID), 129
  Arc/Info surface modeling extensions: (TIN), 129
  cartographic publishing tools: (ARCPLOT), 129
  data conversion: (ARC), 129
  data display and analysis tools: (ARCPLOT), 129

data entry and edit tools: (ARCEDIT), 129
display speed enhancement extensions: (ArcExpress), 129
GIS graphic user interface: (Arc-Tools), 129
geographic data management: (LIBRARIAN), 129
image display and management: (IMAGE INTEGRATOR), 129
interactive graphic form menu creation: (FormEdit), 129
linear modeling: (Dynamic Segmentation), 129
macro language: (AML), 129
map projections, manipulation and management: (ARC), 129
overlapping polygons and area modeling: (Regions), 129
plotter support: (ARCPLOT), 129
polygon overlay and buffering: (ARCPLOT), 129
RDBMS Access: (DATABASE INTEGRATOR), 129
scanning data entry extensions: (ArcScan), 129
spatial data analysis: (ARCPLOT), 129
tabular data management: (INFO), 129
ARC-NET (Australia), 60
ASCII (American Standard Code for Information Interchange), 26, 47
ATLAS, 57
attribute database design, 10, 19
attribute data conversion, 22, 26
attribute data type, 4, 21–2
  descriptive data, 4
  qualitative data, 4
  quantitative data, 4
  thematic data, 4
attribute title, 22
attributes, 4
  linkage between spatial data and, 22

Australian Bureau of Statistics (ABS), 117
Autocad based GIS, 47, 95
automated mapping/facilities management (AM/FM), 11

B
BAPPEDA, 104
  Integrated Land Resources Database (ILRDB), 105
base map data, 59
base maps, 5
benchmark test, 87
Berthier, Louis Alexander, 7
Brisbane City Council, 53
browsing, 26
Brundtland Report, 50
buffer generation, 31

C
cadastral survey, 5
cadastre, 11
CAD/CAM, 74–5
Canadian Geography Information System (CGIS), 8
Candata, 120–1
cartographer, 46
cartographic elements, 35
CD-ROM (compact disc read-only memory), 39
census, 6
central processing unit (CPU), 40
centralized computers, 40
centralized model system architecture, 40
character recognition, 94
classical planning process, 49–50
  adoption of preferred plans and/or policies, 49
  data collection and analysis, 49
  development of alternative plans and/or policies, 49
  evaluation of alternatives, 49
  feedback, 50
  goal setting, 49
  implementation of plans and/or policies, 49
  monitoring and evaluation of the results, 50
  problem identification, 49
  refinement of goals, 49
classification and analysis, 33
  density slicing, 33
  principle component analysis, 33
computer, 18, 35
  configuration, 37

computer-assisted project management (CAPM), 65
  software, 67
conceptual database design, 76
connectivity and contiguity analysis, 31
contrast stretching, 33
control points, establishing the, 93
Coordinate Geometry (COGO), 21
Coordinates/Projection change, 24
cost–benefit analysis, 81
coverage automation, 94

D
data
  acquisition and collection, 23–4
  needs interviews, 74
  presentation, 22, 33, 38
database, 18
  administrator, 46
  editing, 27–8
  integration/management expert, 91
  manipulation, 28, 99
    join and extract, 28
    reclassify, 28
    sort and index, 28
  modeling, 19
  query, 29–30
    basic spatial query, 29
    patterns, 30
    'show me', 30
    simulation, 30
    structured query,
    'tell me', 29
    trends, 30
  structure, 19
    flat file, 19
    hierarchial file, 19
    networks structure, 19
  system, 2
  viewing, 26–7
    browsing, 26
    multiwindow, 27
    panning, 27
    zooming, 27
database management system (DBMS), 8, 22, 26, 69
data capture, 46–7, 77, 90, 92–3
  contract, 92–3
  industry, 78
  specification, 92
data collection, 22
Data Exchange Format (DXF), 47, 49
data exchange standards, 42–3
Data Import, 94
data import, 23–4
  ASCII text, 23

global positioning system (GPS), 23
    raster RS data, 23
    soft photogrammetry data (air photo scan), 23
    total station data (ASCII), 23
Datamap, 54
data modeling, 114
data-processing system, 2
data representation, 48
data sets, 9
data transmission, 1
DB2, 47
Dbase, 47
decision support system (DSS), 2
    graphics, 2
    mapping, 2
    modeling, 2
    query functions, 2
    spreadsheets, 2
density slicing analysis, 33
Department of Administrative Services/Australian Survey and Land Information Group (AUSLIG), 118
description of final products, 93
descriptive data, 4
development tools, 35
Digital Cadastral Data Base (DCDB), 11
    large scale, 53
Digital Elevation Model (DEM), 47, 95
Digital Equipment Corporation (DEC), 37
digital image, 14
    analysis, 48
digital recording, 1
digital spatial data, 16
Digital Terrain Model (DTM), 21
digitally processed information, 1
    see also information technology
digitizer/scanner, 23, 38
digitizing, 38, 94
direct thermal plotter, 39
Disk Operating System (DOS), 47
distributed model, 40
distributed model computer, 40–2
DLG, 95

E
Earth Resources Mapper (ER Mapper), 132
Edge-matching, 24
electromagnetic radiation spectrum, 12
Electronic Numerical Integrator and Calculator (ENIAC), 7
electrostatic plotter, 39–40

emergency planning and management system, 56
end user, 46
ENIAC *see* Electronic Numerical Integrator and Calculator
environmental accounting, 55
environmental database, 55
Environmental Impact Assessment, 55
environmental monitoring, 55
environmental planning, 54
environment risk management, 55
environmental standards, 55
Environmental Systems Research Institute (ESRI), 129
ERDAS, 54, 57–8
ER Mapper *see* Earth Resources Mapper
Ethernet, 44
executive support, 70
executive steering committee, 70
expert system, 2
    inference rules, 2
    knowledge base, 2

F
FASTPAK, 45
Fiber Distributed Data Interface (FDDI), 44
field survey, 77
floppy disk, 40
format conversion, 23
Foxbase, 47
friendly user interface, 47

G
GDS, 54, 59–60
GenaMap, 47, 54–5, 58, 60, 124
Genasys, 54, 124
GenaVive, 124
generalization, 34
    annotation, 34
    cartographic elements, 35
    classification, 34
    combination, 34
    customization of symbols, 35
    display and viewing of raster and vector data, 35
    exaggeration, 34
    selection, 34
    simplification, 34
    symbolization, 34
Genius (Graphic User Interface), 124
geographic data, 3
Geographic Base File/Dual Independent Map Encoding (GBF/DIME), 95

geographic identities, 6
  see also spatial data elements
geographic information, 3–7
  areal extent, 4
  attribute data, 4, 18
  concepts of, 4–5
  human factors, 5
    constructed, 5
    cultural, 5
    economic, 5
    historical, 5
    political, 5
    population, 5
    social, 5
  line, 3
  manipulation of, 6
  needing study, 74
  physical data for, 4–5
  point, 3
  regional data for, 4–5
  spatial data, 3–4, 18
  surveys
    for mapping and cartography, 5
    to obtain, 4–5
      aerial survey/Remote Sensing,
        5–6
      census, 6
      field/hydrographic/mining, 5
        cadastral, 5
        engineering, 5
        geodectic, 5
        topographical, 5
    statistics, 6
    tracking and monitoring, 6
  time, 4
geographic information products
    analysis, 74
geographic information systems *see*
    GIS
georelated database, 19, 22, 26
  modeling, 19
Georelational geographic information
    database management system, 19
Geo/SQL, 58
Geovision, 60
GIS
  acquisition of, 65
  application of, 51
    analysis, 51
    inventory, 51
    modeling, 51
    presentation, 51
  application to planning, 51–2
    database management, 8
    data collection, 8
    digital mapping, 8, 33
    functional components of, 9

configurations, 76
    level of user analytical support,
      76
  components, 18
  data and data collection, 18, 23
    integration approach, 10
    modeling approach, 10
    software, 18
    transformational approach, 9
  database update, 99
  digital mapping of, 8, 33
  diversity of purpose, 11
    automated mapping/facilities
      management (AM/FM), 11
    cadastre (Land Information
      System), 11
    emergency management, 12
    environmental planning, 12
    marine exploration, 11
    modeling systems, 11
    natural resource information
      system, 11
    petroleum exploration, 11
    transportation routing, 11
    urban information system, 11
    weather forecasting, 11
  end-product characteristics, 75
  functionality of, 47
  identifying applications of, 75
  implementation, 63
    multistage approach, 64
    project-driven approach, 63
    supplier–partnership approach, 64
  industry, 123 ff.
  information technology for,
    analysis, 7–8
    collection of data, 8
    management, 8
    representation, 8
  integrated system for, 16
  integration approach, 10
  management unit of, 90
  manipulation
    merits of for planning and
      decision making, 17
    modeling approach, 10
  output capability, 75
  products, 33, 75
    tailored, 75
    user applications, 75
  project implementation, 65
  real-world definition, 11
  resource and environment
      based, 50
  support analytical needs, 76
  technology,
    maintenance, 101

technology acquisition management
    framework, 67
    feasibility study, 67
    prefeasibility investigation, 67
    system development, 67
    system selection, 67
GIS data, marketing of, 100–1
    versatility of, 47
global positioning system (GPS), 47,
    57
Graphic User Interface (GUI), 23
    button-items, 23
    choice-items, 23
    label-items, 23
    pop-up menus, 23
    sliders, 23
    text-items, 23
    toggle-items, 23

H

hard disk, 40
hardware, components, 36
high-resolution visible (HVR) imaging,
    14
human geographic information, 5

I

IDRISI, with Clark University, 133
    anisotropic cost-surface generation,
        133
    bayesian uncertainty analysis, 133
    Fuzzy Set analysis, 133
    multicriteria/multiobjective decision
        making, 133
    tools for advanced analysis and
        modeling, 133
image classification, 15
image enhancement, 15, 33
    contrast stretching, 33
    random noise elimination, 33
    spatial filtering, 33
image processing and preprocessing
    systems, 8, 14–15, 32
    geometric correction, 32
    radiometric correction, 32
    restoration, 32
incremental capture, 92
information
    composition entry from more than
        one data item, 2
    exchange and services, 1
information services
    generation of employment, 1
information superhighway, 1
information system, application of, 2
    data-processing system, 2
    decision support system, 2

    expert system, 2
    geoinformation system, 2
    information-processing system, 2
    integrated information system, 2
    management information system, 2
    operational system, 2
    synonomous with database system,
        2
information technology, 1, 11
    census data-processing systems, 3
    customized management, 3
    decision support system, 3
    expert system, 3
    geographic information system, 3
    statistics analysis packages, 3
information technology project man-
    agement, 66
infrared light (IR), 12
in-house GIS data capture, 77
Initial Graphic Exchange Specification
    (IGES), 47, 95
inkjet/bubblejet plotter, 39
inkjet/bubblejet printer, 39
input device: digitizer/scanner, 38
integrated database management soft-
    ware, 18
integrated databases, 10
integrated information system, 2
Integrated Services Digital Network
    (ISDN), 45
integration approach, 10
interactive display function, 33
interactive voice transmission, 33
Intergraph Modular GIS Environment
    (MGE), 47, 54, 60, 126
International Standards Organization
    (ISO), 45
interpretation mechanism, 15

L

LAN *see* local area network
Lands Department of Guandong
    Province, 110
Landsat, 13, 15
laser printer/plotter, 38
LIC Project Management Methodol-
    ogy (LICPMM), 112
linescan systems, 13
lithographic techniques, 7
    see also map products
local area network (LAN), 40, 44
local government, 52

M

Macintosh, 47
magnetic disk, 40
magnetic tape, 40

mainframe/minicomputer, 36
management information system
    (MIS), 2
manager, 46
manipulating database, 28
  join and extract, 28
  reclassify, 28
  sort and index, 28
manipulation of geographic informa-
    tion, 6
manual mapping system, 8
map
  function of,
    determination of their location, 7
    identification of spatial data ele-
      ments, 7
    measurement of spatial attributes,
      7
    portrayal of data elements, 7
    storage of data elements, 7
map dissolve, 31
  objective of, 33
    area, 34
    attribute, 34
    legends, 34
    projection, 34
    scale, 34
    spatial data, 34
    titles, 34
    units, 34
Map Overlay and Statistical System
    (MOSS), 95
map overlays, hinged, 7
map/polygon analysis function, 47–8
map products, 7
map representation, 8
map scale issues, 93
MapGrafix, 54
MapInfo, 47, 54, 57, 59, 61–2, 131
  GIS software, 131
    buffer generation, 131
    features drawing and data presen-
      tation, 131
    fully functional relational data-
      base with SQL capabilities
      and spatial extensions, 131
    geocoding, 131
    MapBasic, 131
    overlay, 131
    project transformation, 131
    thematic mapping, 131
    vector and raster images display,
      131
  Mapper (Earth Resources Mapping),
    132
  applications of, 132
  DTM editing in 3-D stereo, 132

  DTM generation from stereo pairs,
    132
  feature editing in 3-D stereo, 132
  image orthorectification, 132
  interactive image processing, 132
mapping
  thematic, history of, 7
mapping technology, 6
  analysis, 6
  graphic representation of GIS, 6
marketing GIS data, 100–1
matrix printer, 38
Mercator projection, 25
metadata, 98
MGE Analyst, 126
MGE Dynamic Analysis, 126
MGE Environment Manager, 127
MGE GeoData Manager, 126
MGE Grid Analyst, 127
MGE Network Modeler, 127
MGE Parcel Manager, 126
MGE Projection Manager, 126
MGE Segment Manager, 127
MGE Terrain Modeler, 127
MGE Voxel Analyst, 127
MicroBrian, 124
MicroImages, 125
Microstation, 126
Miller Cylindrical projection, 26
mineral resource management, 57–8
modeling approach, 10
multichannel data transformation and
    exchange, 47
multiple attribute polygon, 31
multistage approach, 64
multiwindow, 27

N
National Map Accuracy Standards,
    96
Native Language Support (NLS), 47
natural resource manager, 16, 54
natural resources information system,
    54
Navigate, 121
near-infrared (NIR) radiation, 12
nearest neighbour search, 31
network adjacency and proximity
    analysis, 48
network analysis, 31, 48
  communications, 32
  geocoding, 32
  roads, rail, canals, airways, 32
  utilities, 32
network protocols, 44
networking standards, 43

New York Landuse and Natural
      Resources Information System, 8

O

Openlook, 43
open-system approach, 41
Open Systems Interconnection (OSI)
      Reference Model, 45
operating system, 2, 37, 43
   DOS (disk operating system), 37, 43
   UNIX, 37, 43
   X-Windows (X-11), 37, 43
optical disk, 39–40
optical scanning, 38
Oracle, 47
orthographic projection, 24
OSF (Open Systems Foundation)
      Motif, 43
OS/2, 43

P

packets, 43–4
Panning, 27
PC Arc/Info, 54
PC Arc/Info Version 3.4D Plus, 106
PC-ArcView, 61
pen plotter, 39
personal computers (PCs), 36–7
   vs mainframe and minicomputer,
      36–7
phased capture, 92
photogrammetry, 6
photographic systems, 12
physical geographic information, 5
picture elements (pixels), 14
pilot project, 89
pilot system development and training
      program, in Zhuhai, 115
pixels, 14
planning, 50
   economic development, 50
   environmental protection, 50
   infrastructure provision, 50
   resource planning, 50
   social welfare improvement, 50
Platforms (satellites), 13
polygon overlaying operations, 30
pop-up menus, 23
potential users, 46
prefeasibility investigation, 67
preprocessing of images, 14
price policy, 100
principal component analysis, 33
printer, inkjet/bubblejet, 39
printer/plotter, laser, 38
production manager, 91
programmer, 46

project-driven approach, 63
proximity analysis, 48
Public Sector Mapping Agency
      (PSMA), 118

Q

QA coordinator, 46
QA/QC and data acceptance staff, 91
   acceptance plans, 91
   quality control procedures, 91
quantitative analysis, land use, 7
quantitative data, 8
querying database, 29–30
   basic spatial query, 29
   patterns, 30
   'show me', 30
   simulation, 30
   Structured Query Language (SQL),
      29
   'tell me', 29
questionnaire survey, 73

R

radar systems, 13
radiation spectrum, electromagnetic,
      12
radiometric corrections, 32
random access memory (RAM),
      35–7
random noise elimination, 33
raster, 22
raster capture, 92
raster data, 21
raster images, 23
raster structures, 13, 21
   data, 94
rasterization, 23, 31
Rbase, 47
real-time monitoring, 32
reduced instruction set computer
      (RISC), 43
regional geographic information, 4
relational database, 19
   modeling, 19
   tables, 19
relational database management
      system (RDBMS), 47, 69
Relational Interface System (RIS),
      126
remote sensing, 6, 8, 9, 12
remotely sensed data, 13, 16, 77
Request for Tender (RFT), 84
   preparation of, 84
restoration, 32
Robinson projection, 25
routing, 32
rubber sheet stretching, 24

S

satellites, 13
scale change, 24
scale of remote sensing data, 16
scanners, optical, 38
scanning, 23
scrub/records preparation staff, 91
seamless base maps, 24
Semarang, Indonesia, 103–8
seminars, to introduce GIS project, 71
sensors, 12
Small Format Aerial Photography
    (SFAP), 16, 77
small-format aerial photogrammetry,
    77
Smallworld GIS, 130
    application configuration environ-
        ment, 131
    CASE tool, 131
    integrated raster/vector/object data-
        base, 131
    topologically structured geometry,
        131
social survey, 6
soft photogrammetry, 23
source data error, 96
SPANS (Spatial Analysis System), 47,
    61–2, 127–8
SPANS Data Translation, 128
SPANS GIS, 127
SPANS GIS Builder I, 128
SPANS GIS Builder II, 128
SPANS MAP, 127–8
SPANS TYDIG, 128
spatial analysis
    map overlays, 7
        network analysis, 7
        surface analysis, 7
spatial data elements, 7
spatial data (GIS), 3, 20, 53–4, 61
    linkage between attributes and, 22
spatial data identifier, 20
spatial data modeling, 19
Spatial Data Transfer Standard
    (SDTS), 47
spatial display techniques, 16
spatial filtering, 33
spatial information system, 10
spatial software model, 16
spectral bands of emissions, 12
speech recognition, 95
SPOT, 13–14
stand-alone, single-user model of
    computer, 40
standards
    data exchange, 42–3
    software (user interface), 43

statistics, 6
strategic planning to acquire a GIS,
    65–6, 70
supercomputer, 36
super-operator, 46
supervised classification, 33
supplier–partnership approach, 64
surface analysis, quantifying the
        components, 32, 48
    area and length calculation, 32
    generation of cross/long sections, 32
    interpolate elevations, 32
    specify internal contouring, 32
    surface slope aspect, 32
    3-D representation, 32
    volumetric cut and fill calculations,
        32
surface representation in 3-D, 21
survey methods, 6
Sybase, 47
symbol recognition, 94
symbolization, 34
symbols, basic, customization of, 6
system, 2
    input data, 2
    manage data, 2
    output data, 2
    process data, 2
    store data, 2
system administrator, 46
system development, 88
system development life cycle
        (SDLC), 66
system overview, 85
system requirements, 85
system selection, 83

T

tabular digitizer, 23
TACTICAN, 62
Tag Image File Format (TIFF), 95
technical committee, 71
tendering conditions, 85
thematic mapping, 7
    history of, 7
themes, 4
thermal plotter, 39
3-D, 14
    imaging, 14
    surface representation, 21
TIGER, 95
time, 4
TNTatlas, 125
TNTdemo, 125
TNTlink, 125
TNTmips (Map and Image Process-
        ing System), 124–5

TNTsdk, 125
TNTview, 125
token ring, 44
topological information, construction of, 24
topological vector data, 21
total capture, 92
Total Quality Management (TQM), 97
Town and Country Planning Act 1947, UK, 7
Tracing algorithms: production-oriented raster to vector conversion, 23, 32
tracking/monitoring, 6
training framework, 79
training needs analysis, 78
training needs matrices, 79–80
transport, 58–9
  typical applications, 58–9
transport demand modeling, 59
transport routing, 11
Transverse Mercator projection, 25
Triangulated Irregular Network (TIN), 21, 47
'turnkey' technology, 18
TYDAC Technologies Inc., 127

U
United Nations Development Program (UNDP), 9
United Nations Economic and Social Commission for Asia and Pacific (UNESCAP), 9
United Nations Environmental Program (UNEP), 9
Universal Transverse Mercator (UTM) projection, 26
UNIX, 43, 47

unsupervised classification, 33
Urban and Regional Information Systems Association, 8
user requirements analysis (URAL), 72–8, 83

V
vector, 21
vector data, 21
vector structure data, 21, 94
vectorization, 22, 23
VGA (Video Graphic Adapter) board, 37
viewing database, 26–7
  browsing, 26
  multiwindow, 27
  panning, 27
  zooming, 27
Virtual Memory System (VMS), 47
visible light spectrum, 12
visualization techniques, 16
voice data entry, 95

W
wavelengths (spectral bands), 12
wide area network (WAN), 40, 44
Windows, 37
workstation, 36
WORM (write once read many), 39
write once read many (WORM), 39

X
X-Terminal, 36, 41
X-Windows (X–11), 37

Z
Zhuhai, Guangdong Province, People's Republic of China, 109
  municipal government, 110
zooming, 27